Progress in Probability
Volume 50

Stochastic Analysis and Mathematical Physics

A.B. Cruzeiro
J.-C. Zambrini
Editors

Springer Science+Business Media, LLC

A.B. Cruzeiro
Grupo de Fisica-Matemática
Universidade de Lisboa
1649-003 Lisboa
Portugal

J.-C. Zambrini
Grupo de Fisica-Matemática
Universidade de Lisboa
1649-003 Lisboa
Portugal

Library of Congress Cataloging-in-Publication Data

A CIP catalogue record for this book is available from the Library of Congress, Washington D.C., USA.

AMS Subject Classifications: 60H30, 60H15, 60H07, 58B20, 58J65, 81S25, 81S99

Printed on acid-free paper

Originally published by Birkhäuser Boston in 2001
Softcover reprint of the hardcover 1st edition 2001

SPIN 10837140
ISBN 978-1-4612-6624-2 ISBN 978-1-4612-0127-4 (eBook)
DOI 10.1007/978-1-4612-0127-4

Reformatted from editors' files by TEXniques, Inc., Cambridge, MA

9 8 7 6 5 4 3 2 1

Contents

Functorial Analysis in Geometric Probability Theory
by Hélène Airault and Paul Malliavin **1**

Stochastic Volterra Equations with Singular Kernels
by L. Coutin and L. Decreusefond **39**

Stochastic Diffeology and Homotopy
by Rémi Léandre **51**

Some Results on Entropic Projections
by C. Léonard **59**

Mehler-Type Semigroups on Hilbert Spaces and Their Generators
by Paul Lescot **75**

Singular Limiting Behavior in Nonlinear Stochastic Wave Equations
by Michael Oberguggenberger and Francesco Russo **87**

Complete Positivity and Open Quantum Systems
by Rolando Rebolledo **101**

Properties of Measure-preserving Shifts on the Wiener Space
by A.S. Üstünel **133**

Martingale and Markov Uniqueness
of Infinite Dimensional Nelson Diffusions
by Liming Wu **139**

Preface

This volume consists of articles which are an outgrowth of an international meeting organized in Lisbon by the Group of Mathematical Physics (GFM). The meeting focused on the subject of stochastic analysis and some of its contact points with mathematical physics.

L. Coutin and L. Decreusefond consider stochastic versions of Voltera equations with singular kernels. The area of nonlinear stochastic wave equations and their singular limit behaviour is investigated by M. Oberguggenberger and F. Russo. The interest of Mehler-type semigroups on Hilbert spaces has its origin in quantum physics and their generators are studied by P. Lescot. Some results about entropic projections, presented by C. Léonard, are also motivated by various probabilistic approaches to quantum theory, whose common origin is a work of E. Nelson, in the mid-1960s. The same idea is also explored in the infinite dimensional case from the point of view of the martingale problem and Markov uniqueness by L. Wu.

Another perspective on the relations between quantum (open) systems and probability theory is described by R. Rebolledo. This requires the concept of complete positivity. In recent years, the geometry of the Wiener space has become an important field of research in stochastic analysis. H. Airault and P. Malliavin present an analysis in geometric probability theory in which the image of a connection is defined by a functor, and they study the properties of the new resulting connection. A.S. Üstünel considers the properties of measures preserving shifts on the Wiener space. Motivated partly by recent developments in mathematical physics, R. Léandre describes a stochastic version of cohomology on the loop space, and groups that are invariant under differentiable homotopies on the manifold.

This research group has received grants from the Portuguese Foundation for Science and Technology (FCT) and organizes periodically international conferences in various fields. For this meeting, the GFM was also sponsored by the French Ambassy and by the Calouste Gulbenkian Foundation. It is a pleasure for the Editors to thank the FCT and the other sponsors for their support, as well as the participants for their active contribution to the Lisbon meeting (May 1999) which was quite enjoyable for everyone.

Ana Bela Cruzeiro
Jean-Claude Zambrini
Lisbon, February 2001

Functorial Analysis in Geometric Probability Theory

Hélène Airault and Paul Malliavin

ABSTRACT We define the image of a connection with the help of a functor. This image is a connection; we study its properties and also the stability of various Riemannian formulae through this functor.

The subject of stochastic calculus of variations is a mixture of probability theory and differential geometry. In classical differential geometry the good functors are the *inverse image* functors. For instance, the inverse image of a differential form is well defined for any smooth map; on the contrary, the direct image of a vector field is defined only for a smooth diffeomorphism. In this article, using conditional expectation, we construct a theory of direct image for the stochastic calculus of variations. We emphasize the algebraic mechanism of these constructions; for this reason, this article is written in the finite dimensional case. We postpone implementing these ideas for probability spaces of infinite dimension to another work.

Notation. A *regular probability space* will consist of the data of a smooth finite dimensional Riemannian manifold M, with Riemannian connection, together with an absolutely continuous probability measure μ, and having smooth density relative to the Riemannian volume measure dv of M. *A regular morphism between two regular probability spaces M* and $\widetilde{M}$ will be a smooth map $g : M \to \widetilde{M}$, such that for any $m \in M$, its differential $g'(m)$ is surjective. Then the image measure $\widetilde{\mu} = g_*\mu$ is absolutely continuous with respect to the Riemannian volume $d\widetilde{v}$ on $\widetilde{M}$. Assuming that g is smooth and $g'(m)$ is surjective, the measure μ disintegrates continuously along g and the conditional expectation E^g defines a map from $C^k(M)$ to $C^k(\widetilde{M})$. We can define the direct image of a function only if there exists a measure μ on M.

In Part 0, we list the main formulae involving direct images and inverse images. In Part I, we present geometric tools and we give their expression in local coordinates. We extend the divergence formula in terms of the trace of covariant derivatives to the case of a non-Riemannian connection on M. See (0.27)–(0.28).

In Part II, let g be a differentiable map from the manifold M to the manifold $\widetilde{M}$; we develop the theory of image. The functor inverse image g^* is classical in

---

AMS subject classification 60H07

differential geometry. With probability measure μ on M, we define the functor direct image g_*. We study functoriality properties of geometric identities through g^* and g_*. We define a metric g_*a on $\tilde{M}$ from a metric a on M: For two vector fields $\tilde{Y}$ and $\tilde{Z}$ on $\tilde{M}$, we define the metric on $\tilde{M}$ with $(\tilde{Y}|\tilde{Z})_{g_*a} = g_*((g^*(\tilde{Y})|g^*(\tilde{Z}))_a)$. This construction differs from that of [1] in the sense that in [1], the image metric is defined using the gradient ∇_a associated to the metric a on M. We give a conditional requirement so that the image metric g_*a coincides with that of [1]. In the same way we define the image of a tensor. We obtain a tensor on $\tilde{M}$. With (0.19), we define the image of a connection. This definition does not require the map g to be a diffeomorphism (compare with [4], p. 79). However, as above, we need a measure on the manifold M. The image of a connection is a connection. However, consider a Riemannian connection on M (i.e., parallel transport is an isometry), then its image is not necessarily a Riemannian connection for the image metric. An example is given in Part II, 7. In Part III, following Sasaki [6], we extend the Riemannian geometry to the tangent bundle $T(M)$ of the Riemannian manifold M. We give on the tangent bundle a brief survey of our method of direct and inverse images.

Part 0 The functors g_* and g^*

This section is a compilation of formulae involving the functors g_* and g^*, which will be proved in Parts I and II. M is a Riemannian manifold endowed with the Riemannian metric a, μ is a measure on M such that μ is absolutely continuous with respect to the volume dv on M. Let $\tilde{M}$ be a differentiable manifold and consider a smooth map $g : M \to \tilde{M}$, such that for any $m \in M$, its differential $g'(m)$ is surjective. We introduce the operator g^* on functions and on differential forms by the classical formulas

$$g^*\tilde{f} = \tilde{f}og \quad \text{and} \quad (g^*\tilde{\omega})_x = ({}^t g'(x))(\tilde{\omega}_{g(x)}) \tag{0.1}$$

We split the tangent space $T_m(M)$ of the Riemannian manifold M as the direct sum of $\ker(g'(m))$ with its orthogonal; then the restriction of $g'(m)$ to its orthogonal is an isomorphism; we denote by $(g'(m))^{-1}$ its inverse. Let a vector field $\tilde{Z}$ be given on $\tilde{M}$. We define

$$(g^*(\tilde{Z}))_m = (g'(m))^{-1}(\tilde{Z}_{g(m)}) \tag{0.2}$$

We have

$$< d(g^*\tilde{f}), g^*\tilde{Z} >= g^*(< d\tilde{f}, \tilde{Z} >) \tag{0.3}$$

By using conditional expectation, see [5 p. 76–82], we construct operators of direct image, pushing down functions and vector fields from M to $\tilde{M}$ by

$$g_*f = E^g_\mu(f) \quad \text{and} \quad (g_*(Z))(\tilde{f}) = E^g_\mu(Z(\tilde{f}og)) \tag{0.4}$$

Thus, the functor g_* depends on the measure μ on M, while the functor g^* depends on the Riemannian metric a on M. In the following, the measure μ and the Riemannian metric on M are fixed, and we do not specify at each step that g_* depends on μ. For $f : M \to R$ and $\widetilde{\phi} : \widetilde{M} \to R$, it holds that

$$g_*((g^*\widetilde{\phi})f) = \widetilde{\phi}g_*(f) \tag{0.5}$$

For the functors g^* and g_*, we have

$$g_*g^* = Id_{\text{functions on}\widetilde{M}} \quad \text{but} \quad g^*g_* \neq Id_{\text{functions on}M} \tag{0.6}$$

$$g_*g^* = Id_{\text{vector fields on}\widetilde{M}} \quad \text{but} \quad g^*g_* \neq Id_{\text{vector fields on}M} \tag{0.7}$$

For the Lie bracket of two vector fields on $\widetilde{M}$, we have

$$g_*[g^*(\widetilde{Z}_1), g^*(\widetilde{Z}_2)] = [\widetilde{Z}_1, \widetilde{Z}_2] \quad \text{but} \quad [g^*(\widetilde{Z}_1), g^*(\widetilde{Z}_2)] \neq g^*[\widetilde{Z}_1, \widetilde{Z}_2] \tag{0.8}$$

If $\widetilde{c}$ is a chart on $\widetilde{M}$, and $(Z_j^{\widetilde{c}})$ is the local basis in the chart $\widetilde{c}$, then

$$g_*(Z) = \sum_j g_*(Z(g^*\widetilde{c}_j))Z_j^{\widetilde{c}} \tag{0.9}$$

$$div_{g_*\mu}(g_*Z) = g_*(div_\mu Z) \quad \text{and} \quad div_{g_*\mu}(\widetilde{Y}) = g_*(div_\mu(g^*(\widetilde{Y}))). \tag{0.10}$$

Let f be a function on M and let $\widetilde{Z}$ be a vector field on $\widetilde{M}$. We have

$$g_*(fg^*(\widetilde{Z})) = (g_*f)\widetilde{Z} \tag{0.11$_1$}$$

and (see [5 p. 82])

$$\widetilde{Z}(g_*f) - g_*\big((g^*\widetilde{Z})f = f \quad div_\mu(g^*\widetilde{Z})\big) - (g_*f)g_*\big(div_\mu(g^*\widetilde{Z})\big). \tag{0.11$_2$}$$

For a vector field Y on M, we may have $g_*(Y) = 0$ but $g_*(fY) \neq 0$.

The method extends pushing down from M to $\widetilde{M}$ differential forms and tensor fields. For a differential form ω of degree one on M and a vector field $\widetilde{Z}$ on $\widetilde{M}$, we put

$$< g_*(\omega), \widetilde{Z} >= g_*(< \omega, g^*\widetilde{Z} >). \tag{0.12}$$

Then

$$g_*g^* = Id_{\text{differential forms on}\,\widetilde{M}} \quad \text{but} \quad g^*g_* \neq Id_{\text{differential forms on}\,M}. \tag{0.13}$$

For a function $f : M \to R$, we denote by $\nabla_a f$ the gradient associated to the Riemannian metric a on M. In [1], the image metric g_*a on $\widetilde{M}$ was such that

$$\nabla_{g_*a}\widetilde{\phi} = g_*(\nabla_a(g^*\widetilde{\phi})) \tag{0.14}$$

where ∇_{g_*a} denotes the gradient assocated to the metric g_*a on $\widetilde{M}$. In the following, we denote by $(\quad|\quad)_b$ the scalar product in the tangent bundle of a Riemannian manifold endowed with a Riemannian metric $b = b_{ij}$. We define a metric $b = g_*a$ on $\widetilde{M}$ with

$$(\widetilde{Y}|\widetilde{Z})_{g_*a} = g_*((g^*(\widetilde{Y})|g^*(\widetilde{Z}))_a) \tag{0.15}$$

where $\widetilde{Y}$ and $\widetilde{Z}$ are two vector fields on $\widetilde{M}$. The metric $b = g_*a$ depends on the measure μ on the manifold M. For two vector fields Y and Z on M, in general

$$g_*((Y|Z)_a) \neq (g_*(Y)|g_*(Z))_{g_*a}. \tag{0.16}$$

We compare the image metric defined with (0.15) with that of [1], which satisfies (0.14). The method extends to tensors. For example, let ψ be a tensor of type (0.2) on M with real values. We define the image tensor $g_*\psi$ on the manifold $\widetilde{M}$ by

$$(g_*\psi)(\widetilde{Y}, \widetilde{Z}) = g_*(\psi(g^*\widetilde{Y}, g^*\widetilde{Z})) \qquad (0.17)_1$$

The image tensor $g_*\psi$ depends on μ through g_* and on the metric a through g^*. Because of $g^*g_* \neq Id_{\text{vector fields on } M}$, see (0.7), we have in general that

$$g_*(\psi(Y, Z)) \neq (g_*\psi)(g_*Y, g_*Z) \qquad (0.17)_2$$

Let $\widetilde{f}$ be a function on $\widetilde{M}$. To define the image connection, we use the following properties:

$$g^*(\widetilde{f}\widetilde{Y}) = g^*(\widetilde{f})g^*(\widetilde{Y}) \qquad (0.18)_1$$

$$g_*(g^*(\widetilde{f})Y) = \widetilde{f}g_*(Y). \qquad (0.18)_2$$

Let ∇ be the covariant derivative associated to a connection on M. We define a connection on $\widetilde{M}$ with its covariant derivative $\widetilde{\nabla} = g_*(\nabla)$:

$$\widetilde{\nabla}_{\widetilde{Y}}\widetilde{Z} = g_*(\nabla_{g^*(\widetilde{Y})}(g^*(\widetilde{Z}))) \tag{0.19}$$

In general

$$g^*(\widetilde{\nabla}_{\widetilde{Y}}\widetilde{Z}) \neq \nabla_{g^*(\widetilde{Y})}(g^*(\widetilde{Z})). \tag{0.20}$$

For three vector fields $\widetilde{Y}$, $\widetilde{Z}$ and $\widetilde{U}$ on $\widetilde{M}$, we have

$$(g_*(\nabla)_{\widetilde{Y}}\widetilde{Z}|\widetilde{U})_{g_*a} = g_*((\nabla_{g^*(\widetilde{Y})}g^*(\widetilde{Z})|g^*(\widetilde{U}))_a). \tag{0.21}$$

The image of a Riemannian connection with g_* is not necessarily Riemannian. Let T_∇ be the torsion of the connection ∇ and $T_{g_*(\nabla)}$ the torsion of the image connection $g_*(\nabla)$. Then

$$T_{g_*(\nabla)}(\widetilde{Y}, \widetilde{Z}) = g_*(T_\nabla(g^*\widetilde{Y}, g^*\widetilde{Z})). \tag{0.22}$$

If we define $g_*(T_\nabla)$ in a way similar to that of (0.22) and if we compare with $(0.17)_1$, we see that $g_*(T_\nabla) = T_{g_*(\nabla)}$.

We consider a connection γ on M and we denote ∇_Z the covariant derivative along the vector field Z for this connection. To the connection γ, we associate another connection γ'; the covariant derivative of γ' is defined by

$$\nabla'_Y X = \nabla_X Y - [X, Y] \tag{0.23}$$

The torsion of γ is $T(Y, Z) = \nabla_Y Z - \nabla'_Y Z$. Let c be a chart and $Y = \sum_j (\beta_j o c) Z^c_j$ a vector field on M expressed in the local basis $(Z^c_j)_{j=1,\dots,n}$. We denote $(\nabla_{Z_i} Y)^i$

to be the component of $\nabla_{Z_i} Y$, and $(T(Y, Z_i^c))^i$ the component of $T(Y, Z_i^c)$ along Z_i^c. For the vector field Y and the connection γ, we denote the traces

$$\mathcal{A}_\gamma(Y) = \sum_i (\nabla_{Z_i^c} Y)^i \tag{0.24}$$

$$\mathcal{B}_\gamma(Y) = \sum_i (\nabla_Y Z_i^c)^i + [Z_i^c, Y]c_i \tag{0.25}$$

$$\mathcal{T}_\gamma(Y) = \sum_i (T(Y, Z_i^c))^i = \mathcal{B}_\gamma(Y) - \mathcal{A}_\gamma(Y). \tag{0.26}$$

We verify that $\mathcal{A}$, $\mathcal{B}$ and $\mathcal{T}$ do not depend on the chart c. For the associated connection γ', we have $\mathcal{A}_{\gamma'}(Y) = \mathcal{B}_\gamma(Y)$ and $\mathcal{B}_{\gamma'}(Y) = \mathcal{A}_\gamma(Y)$. If *the connection* γ *is Riemannian*, then the divergence of a vector field on M can be expressed with a trace formula involving covariant derivatives; if γ is Riemannian, we have

$$div_\mu(Y) = Y(\log \frac{d\mu}{dv}) + \mathcal{B}_\gamma(Y). \tag{0.27}$$

However, we found non-Riemannian connections for which (0.27) is true. If *the associated connection* γ' *is Riemannian*, then

$$div_\mu(Y) = Y(\log \frac{d\mu}{dv}) + \mathcal{A}_\gamma(Y) \tag{0.28}$$

Consider the traces $\mathcal{A}$, $\mathcal{B}$ and $\mathcal{T}$, associated to the connection ∇ on M and let $g_*\mathcal{A}$, $g_*\mathcal{B}$ and $g_*\mathcal{T}$ be the traces on $\widetilde{M}$ associated to the image connection $g_*\nabla$; see (0.24)–(0.25)–(0.26), for a vector field Y on M,

$$g_*(\mathcal{T}(Y)) \neq (g_*\mathcal{T})(g_*Y). \tag{0.29}$$

We study in 2 the functoriality of $\mathcal{B}$ and $\mathcal{A}$ under g_* and g^*. We shall see that (0.27) is not functorial under g_*.

Part I The trace formula for covariant derivatives

M is an n-dimensional Riemannian manifold. For $x \in M$, we denote by $T_x(M)$ the tangent space at x. We shall discuss the validity of the identity

$$div_\mu(Y) = Y(\log \frac{d\mu}{dv}) + \sum_i (\nabla_{Z_i^c} Y)^i$$

where Y is a vector field on M, ∇ is the covariant derivative associated to a connection on M and Z_i^c are coordinate vector fields, dv is the volume measure on M and μ is another measure on M. For that purpose, we introduce the notation and geometric tools. In each subsection of this section, there is a trace formula.

1 Vector Fields and Tensors

Let Z be a vector field on M. For any $x \in M$, we have $Z(x) \in T_x(M)$ and

$$Zf(x) = \lim_{t\to 0} \frac{f(\phi_t(x)) - f(x)}{t}$$

where $f : M \to R$ is a differentiable function and $\phi_t(x)$ is a differentiable curve on M satisfying $\phi_0(x) = x$. If M is a vector space, one can take $\phi_t(x) = x + tZ(x)$. Below we write the expression of $Z(x)$ *in local coordinates.* Let c be a chart on M; the chart c is a differentiable diffeomorphism from an open set in M to its image in the space R^n. To simplify notation, we take c to be a diffeomorphism map from M to R^n and we do not care for the domain of definition of the chart c. For a differentiable map $f : M \to R$, we have

$$fo\phi_t(x) = (foc^{-1})(c(\phi_t(x)));$$

thus

$$\begin{aligned}\frac{d}{dt}_{|t=0} fo\phi_t(x) &= d(foc^{-1})(c(\phi_0(x)))\frac{d}{dt}_{|t=0}(co\phi_t(x)) \\ &= \sum_{i=1}^{n} \alpha_i(c(x))\frac{\partial}{\partial c_i}(foc^{-1})(c(x)) = (Zf)(x). \qquad (1.1)\end{aligned}$$

We say that $\alpha_i(c(x))$, $i = 1, .., n$ are the local coordinates of the vector field Z in the chart c.

Let $c(x) = (c_1(x), \dots, c_n(x)) \in R^n$. We have

$$\alpha_i(c(x)) = Zc_i(x). \qquad (1.2)$$

If $\tilde{c}$ is another chart, assume that $(Zf)(x) = \sum_{i=1}^{n} \tilde{\alpha}_i(\tilde{c}(x))\frac{\partial}{\partial \tilde{c}_i}(fo\tilde{c}^{-1})(\tilde{c}(x))$, and then the change of coordinates formula is

$$\alpha_j(c(x)) = \sum_i \tilde{\alpha}_i(\tilde{c}(x))\frac{\partial(co\tilde{c}^{-1})_j}{\partial \tilde{c}_i}(\tilde{c}(x)). \qquad (1.3)$$

For convenience, we abbreviate (1.3) as

$$\alpha_j = \sum_i \tilde{\alpha}_i \frac{\partial c_j}{\partial \tilde{c}_i}. \qquad (1.3)_{bis}$$

We call coordinates vector fields in the chart c, the vector fields defined by

$$(Z_i^c f)(x) = \frac{\partial}{\partial c_i}(foc^{-1})(c(x)) \qquad (1.4)$$

If Z is given by (1.1), we have that

$$Z = \sum_{i=1}^{n} (\alpha_i oc) Z_i^c. \qquad (1.5)$$

Let c and $\widetilde{c}$ be two charts of M; then

$$Z_j^{\widetilde{c}} = \sum_i Z_j^{\widetilde{c}}(c_i) Z_i^c. \tag{1.6}$$

Let dc_i be the coordinate differential forms on M defined by

$$dc_i(x) Z_j^c(x) = \delta_{ij}$$

where $\delta_{ij} = 1$ if $i = j$ and $\delta_{ij} = 0$ if $i \neq j$. The change of coordinates formula for a differential form $\omega(x) = \sum_i \omega_i(x) dc_i(x) = \sum_j \widetilde{\omega}_j(x) d\widetilde{c}_j(x)$ is

$$\omega_i = \widetilde{\omega}_j \frac{\partial \widetilde{c}_j}{\partial c_i} \quad \text{and} \quad d\widetilde{c}_j = \frac{\partial \widetilde{c}_j}{\partial c_k} dc_k \tag{1.3)$_{tierce}$$$

If $Z = \sum_i (\alpha_i o c) Z_i^c$ and $\omega = \sum_j \omega_j dc_j$, then

$$\omega(x) Z(x) = \sum_i (\alpha_i o c)(x) \omega_i(x). \tag{1.7}$$

Let F be a fiber bundle over M and for any $x \in M$, we give a bilinear map $\phi_x : T_x(M) \times T_x(M) \to F_x$; it defines a tensor of type (0,2) with values in F. See [3]. With $Z_j^{\widetilde{c}} = \frac{\partial c_i}{\partial \widetilde{c}_j} Z_i^c$, we deduce the change of coordinates formula

$$\phi(Z_j^{\widetilde{c}}, Z_k^{\widetilde{c}}) = \frac{\partial c_s}{\partial \widetilde{c}_j} \frac{\partial c_p}{\partial \widetilde{c}_k} \phi(Z_s, Z_p). \tag{1.8}$$

If $F = T(M)$, then

$$d\widetilde{c}_i[\phi(Z_j^{\widetilde{c}}, Z_k^{\widetilde{c}})] = \frac{\partial \widetilde{c}_i}{\partial c_p} \frac{\partial c_s}{\partial \widetilde{c}_j} \frac{\partial c_q}{\partial \widetilde{c}_k} dc_p[\phi(Z_s, Z_q)] \tag{1.9}$$

defines a tensor of type (1,2) with real values. The trace

$$\sum_i dc_i[\phi(Z_i^c, Z_k^c)] \tag{1.10}$$

is a tensor of type (0.1).

Let Z and Y be two vector fields with local coordinates $\alpha_i(c(x))$ and $\beta_i(c(x))$. Then

$$(YZ - ZY) f(x) = \sum_{i,j} (\beta_i \frac{\partial}{\partial c_i} \alpha_j - \alpha_i \frac{\partial}{\partial c_i} \beta_j)(c(x)) \frac{\partial (f o c^{-1})}{\partial c_j}(c(x)). \tag{1.11}$$

Let Z be given by (1.1); then the vector field $[Z_k^c, Z] = Z_k^c Z - Z Z_k^c$ is such that

$$[Z_k^c, Z] f(x) = \sum_i \frac{\partial \alpha_i}{\partial c_k}(c(x)) \frac{\partial (f o c^{-1})}{\partial c_i}(c(x)). \tag{1.12}$$

In particular, we have $[Z_k^c, Z]c_k(x) = \frac{\partial \alpha_k}{\partial c_k}(c(x)) = Z_k^c(\alpha_k o c)(x)$. If c and $\widetilde{c}$ are two charts, then

$$\sum_i \frac{\partial}{\partial c_i}(\alpha_i)(c(x)) = \sum_i Z_i^c(\alpha_i o c)(x) = \sum_i (Z_i^c Z - Z Z_i^c)c_i \tag{1.13}$$

$$= \sum_i (Z_i^{\widetilde{c}} Z - Z Z_i^{\widetilde{c}})\widetilde{c}_i + \sum_{i.j} Z(\frac{\partial (co\widetilde{c}^{-1})_i}{\partial \widetilde{c}_j}(\widetilde{c}(x))) \times \frac{\partial (\widetilde{c}oc^{-1})_j}{\partial c_i}(c(x)) \tag{1.14}$$

$$= \sum_i (Z_i^{\widetilde{c}} Z - Z Z_i^{\widetilde{c}})\widetilde{c}_i + Z(\log|\det \frac{\partial (co\widetilde{c}^{-1})_j}{\partial \widetilde{c}_i}(\widetilde{c}(x))|). \tag{1.15}$$

where at the last step, we use the derivation identity for the determinant of a matrix J_{ij}

$$Z(\log|\det J_{ij}|) = -\sum_{ij} J_{ij} Z((J^{-1})_{ji}) = \sum_{ij} (J^{-1})_{ji} Z(J_{ij}). \tag{1.16}$$

From (1.12) we have that for any i, k,

$$Z_k^c Z_i^c - Z_i^c Z_k^c = 0. \tag{1.17}$$

The relation (1.15) is a particular case of the following proposition.

Proposition. *Let* $(v_i)_{i=1,\dots,k}$ *be a set of independent vector fields and put*

$$w_j = \sum_{i=1}^k b^i_j v_i \quad \text{for} \quad j = 1, \dots, k. \tag{1.18}$$

We assume that $\det b^i_j \neq 0$ *and for any* i, j, *the Lie bracket condition holds:*

$$[w_i, w_j] = \sum_{s.p} b_i^s b_j^p [v_s, v_p]. \tag{1.19}$$

Then, for a vector field

$$Z = \sum_i \beta_j w_j = \sum_i \alpha_i v_i, \tag{1.20}$$

we have

$$\sum_j v_j(\alpha_j) = \sum_i w_i(\beta_i) + Z(\log|\det b^j_p|). \tag{1.21}$$

Proof. Since

$$[w_i, w_j] = \sum_k (w_i(b_j^k) - w_j(b_i^k))v_k + \sum_{s.p} b_i^s b_j^p [v_s, v_p],$$

the Lie bracket condition (1.19) is equivalent to the condition

$$w_i(b^j_k) = w_k(b^j_i) \quad \text{for any} \quad i, j, k. \tag{1.22}$$

With (1.17), we see that this condition is well satisfied on an open set of M if we take for (v_i) the coordinate vector fields (Z^c_i) and for (w_j) any system of vector fields satisfying (1.18) and $[w_i, w_j] = 0$ for $i, j = 1, \ldots, k$; for example, we can take for (w_j) the coordinates vector fields $(Z^{\tilde{c}}_i)$. The condition $[w_i, w_j] = 0$ is an integrability condition, see for example the case where $M = R^n$, $p = (p_1, p_2, \ldots, p_n) \in R^n$, $v_i = \frac{\partial}{\partial p_i}$, $i = 1, \ldots, n$, and $w_i = \xi_i(p)\frac{\partial}{\partial p_i}$, $i = 1, \ldots, n$, where $\xi(p) = (\xi_1(p), \xi_2(p), \ldots, \xi_n(p))$ is a function from R^n to R^n.

Corollary. *Let Z, $(v_i)_{i=1,\ldots,k}$ and $(w_j)_{j=1,\ldots,k}$ be vector fields as in (1.18)–(1.19)–(1.20). We assume moreover that, for any i, j, the bracket vector field $[v_i, v_j]$ is in the subfiber bundle generated by the $(v_i)_{i=1,\ldots,k}$. Then*

$$\sum_i [v_i, Z]^i = \sum_j [w_j, Z]^j + Z(\log|\det b^j_p|) \tag{1.23}$$

where $[v_i, Z]^i$ denotes the component of $[v_i, Z]$ along v_i; that is, $[v_i, Z] = \sum_k [v_i, Z]^k v_k$ and $[w_j, Z]^j$ is the component of $[w_j, Z]$ along w_j; i.e., $[w_j, Z] = \sum_k [w_j, Z]^k w_k$.

Proof. This is a consequence of (1.19)–(1.21). Remark that the additional assumption on the Lie bracket of $[v_i, v_j]$ makes (1.23) weaker than (1.21), thus we shall use (1.21) rather than (1.23).

Definition. We say that two system of k independent vector fields $(v_i)_{i=1,\ldots,k}$ and $(w_j)_{j=1,\ldots,k}$ are equivalent if there exists an inversible matrix tensor such that (1.18) and (1.19) hold.

We verify that this defines a relation of equivalence.

Notation. Let ψ be a tensor of type (0,2) on M with value in R. That is, for each $x \in M$, we associate a bilinear form $\psi(x) : T_x(M) \times T_x(M) \to R$. We have

$$\psi(w_i, w_j) = \sum_{s,p} b^s_i b^p_j \psi(v_s, v_p). \tag{1.24}$$

We denote

$$\det_{(v_i)}(\psi) = \det(\psi(v_i, v_j)), \tag{1.25}$$

and we assume that $\det_{(v_i)}(\psi) \neq 0$.

Let $\mathcal{W}$ be a class of equivalence of k independent vector fields; let $(w_j)_{j=1,\ldots,k}$ and $(v_i)_{i=1,\ldots,k}$ be two elements in $\mathcal{W}$ such that (1.18)–(1.19) hold; to a vector field $Z = \sum_i \alpha_i v_i$ in the subspace generated by the $(v_i)_{i=1,\ldots,k}$, we associate the real number

$$Trace_Z^{\psi}(\mathcal{W}) = \sum_i v_i(\alpha_i) + \frac{1}{2} Z(\log |\det_{(v_i)}(\psi)|). \tag{1.26}$$

Lemma. *$Trace_Z^{\psi}(\mathcal{W})$ in (1.26) depends only on Z, ψ and the equivalence class $\mathcal{W}$.*

Proof. By (1.21), we have

$$\sum_i v_i(\alpha_i) - \frac{1}{2} Z(\log |\det b_p^j|) = \sum_i w_i(\beta_i) - \frac{1}{2} Z(\log |\det(b_p^j)^{-1}|).$$

On the other hand, from (1.24)–(1.25),

$$Z(\log |\det \psi(w_i, w_j)|) = 2Z(\log |\det b_s^j|) + Z(\log |\det \psi(v_i, v_j)|).$$

2 Divergence

Let μ be a measure on M and Z be a vector field. Then $div_\mu(Z)$ is a map from M to R such that for any differentiable function $h : M \to R$, we have

$$\int div_\mu(Z)(x)h(x)d\mu(x) = -\int (Zh)(x)d\mu(x). \tag{2.1}$$

For the Lie bracket $[Y, Z]$ of two vector fields Y and Z, we have

$$div_\mu[Y, Z] = Y(div_\mu Z) - Z(div_\mu Y). \tag{2.2}$$

Let μ_1 and μ_2 be two measures on M. If μ_2 is absolutely continuous with respect to μ_1, then

$$div_{\mu_2}(Z) = div_{\mu_1}(Z) + Z(\log \frac{d\mu_2}{d\mu_1}). \tag{2.3}$$

Let c be a chart on M. The image measure $c_*\mu$ is absolutely continuous with respect to dy, the Lebesgue measure on R^n.

Let $(Zf)(x) = \sum_{i=1}^n \alpha_i(c(x)) \frac{\partial}{\partial c_i}(foc^{-1})(c(x))$ be the expression in c of a vector field Z. By (1.4), we have

$$Z_i^c(\alpha_i oc)(x) = \frac{\partial}{\partial c_i}(\alpha_i)(c(x)), \tag{2.4}$$

and we deduce, after integration by parts that

$$(div_\mu Z)(x) = Z((\log \frac{d(c_*\mu)}{dy})oc) + \sum_i Z_i^c(\alpha_i oc)(x). \tag{2.5}$$

See (1.13)–(1.15) for the change of coordinates in $\sum_i Z_i^c(\alpha_i oc)(x)$. Let $d(c_*\mu) = \phi dy$ and $d(\tilde{c}_*\mu) = \tilde{\phi} dy$, we verify that $\frac{\phi oc(x)}{\tilde{\phi} o\tilde{c}(x)} = |\det[\frac{\partial(\tilde{c}oc^{-1})_j}{\partial c_i}]|$. Comparing this identity with (1.15) confirms that $(div_\mu Z)(x)$ in (2.5) does not depend on the chart c.

3 Riemannian metric on M

On each tangent space $T_x(M)$, we have a scalar product; we express this scalar product in a chart c. Consider two vector fields Z_1 and Z_2 and their expression in the chart c:

$$\begin{aligned} Z_1 f(x) &= \sum_{i=1}^{n} \alpha_i(c(x)) \frac{\partial}{\partial c_i}(foc^{-1})(c(x)) \text{ and} \\ Z_2 f(x) &= \sum_{i=1}^{n} \beta_i(c(x)) \frac{\partial}{\partial c_i}(foc^{-1})(c(x)). \end{aligned}$$

The scalar product $(Z_1(x)|Z_2(x))_{T_x(M)}$ is given by

$$(Z_1(x)|Z_2(x))_{T_x(M)} = \sum_{i,j} \alpha_i(c(x))\beta_j(c(x))a_{ij}(c(x)) \tag{3.1}$$

where a_{ij}, which is a map from R^n into the space of real symmetric $n \times n$ matrices, defines the metric in c. Fot the two vector fields Z_1 and Z_2, we also denote $(Z_1|Z_2)_a$ as their scalar product with the metric a. For the metric, the change of coordinates is given by

$$\tilde{a}(\tilde{c}(x)) = {}^t M(\tilde{c}(x))a(c(x))M(\tilde{c}(x)) \qquad \text{where} \qquad M_{jp} = \frac{\partial(co\tilde{c}^{-1})_j}{\partial \tilde{c}_p}. \tag{3.2}$$

The coordinates vector fields $(Z_i^c)_{i=1,\dots,n}$ defined by (1.4) satisfy

$$(Z_i^c|Z_j^c)_{T_x(M)} = a_{ij}(c(x)). \tag{3.3}$$

Since $\det a_{ij} \neq 0$ and any Z given by (1.1) is expressed as $Z = \sum_{i=1}^n (\alpha_i oc) Z_i^c$, then for $x \in M$, the vectors $(Z_i^c(x))_{i=1,\dots,n}$ form a basis of the tangent space $T_x(M)$.

Proposition. *Let $(V_i)_{i=1,\dots,k}$ be k linearly independent vector fields on $T(M)$. We put*

$$b_{ij}(x) = (V_i(x)|V_j(x))_a \tag{3.4}$$

and we define the vector fields

$$W_i = \sum_j b_{ij}^{-1} V_j. \tag{3.5}$$

If $Z = \sum_k \alpha_k V_k$, then

$$\sum_i ([W_i, Z]|V_i)_a + ([V_i, Z]|W_i)_a = \sum_{i,j} b_{ij}^{-1} Z(b_{ij}) + 2\sum_j V_j(\alpha_j). \tag{3.6}$$

4 Riemannian volume

The expression of the Riemannian volume dv in the chart c is the image measure c_*dv. The measure c_*dv is absolutely continuous with respect to the Lebesgue measure dy on R^n and the density is equal to $\sqrt{|\det a_{ij}|}$ where a_{ij} is the Riemannian metric on M. We put

$$\int_M f(x)dv(x) = \int (foc^{-1})(y)\sqrt{|\det a_{ij}(y)|}dy \tag{4.1}$$

and

$$c_*dv = \sqrt{|\det a_{ij}(y)|}dy. \tag{4.2}$$

The definition (4.1) does not depend on the chart c since

$$\int (foc^{-1})(y)\sqrt{|\det a_{ij}(y)|}dy = \int (fo\tilde{c}^{-1})(y)\sqrt{|\det \tilde{a_{ij}}(y)|}dy$$

Let $Zf(x) = \sum_{i=1}^n \alpha_i(c(x))\frac{\partial}{\partial c_i}(foc^{-1})(c(x))$. We deduce from the divergence formula (2.5) and from (4.2) that

$$\begin{aligned} div_{dv}(Z)(x) &= Z(\log \frac{d(c_*dv)}{dy}oc) + \sum_i \frac{\partial}{\partial c_i}(\alpha_i)(c(x)) \\ &= \frac{1}{\sqrt{|\det a_{ij}(c(x))|}} \sum_{i=1}^n \frac{\partial}{\partial y_i}(\alpha_i\sqrt{|\det a_{ij}(c(x))|})(c(x)). \end{aligned} \tag{4.3}$$

Lemma. Let $\nabla_a c_i = \sum_j a_{ij}^{-1} Z_j^c$, see (5.4) below; we have

$$div_{dv}(Z) = \frac{1}{2}\sum_i ([\nabla_a c_i, Z]|Z_i^c)_a + ([Z_i^c, Z]|\nabla_a c_i)_a). \tag{4.4}$$

To prove (4.4), we use (3.3) to calculate

$$\begin{aligned} &([a_{ij}^{-1}Z_j^c, Z]|Z_i^c)_a + ([Z_i^c, Z]|a_{ij}^{-1}Z_j^c)_a \\ &= -a_{ij}Z(a_{ij}^{-1}) + a_{ij}^{-1}\left[([Z_j^c, Z]|Z_i^c)_a + ([Z_i^c, Z]|Z_j^c)_a\right]. \end{aligned}$$

Then, we deduce (4.4) from (4.3) and (1.14)-(1.16). We remark that (4.4) is also a consequence of (3.6).

5 For a function, the gradient vector field associated to the Riemannian metric on M

Let $f : M \to R$ be a differentiable function. The gradient vector field $\nabla_a f(x)$ associated to the Riemannian metric a on M is such that for any vector field Z on M, we have

$$Zf(x) = D_Z f(x) = (\nabla_a f(x)|Z(x))_{T_x(M)} \tag{5.1}$$

The expression (5.1) defines $\nabla_a f(x)$ since $Z(x) \to D_Z f(x)$ is a linear form on a finite dimensional space. Thus there exists a unique $\nabla_a f(x)$ such that (5.1) is satisfied. In local coordinates, let $Zf(x) = \sum_{i=1}^n \alpha_i(c(x))\frac{\partial}{\partial c_i}(foc^{-1})(c(x))$. Then

$$(\nabla_a f(x)|Z(x))_{T_x(M)} = Zf(x) = (\alpha(c(x))|\nabla(foc^{-1})(c(x)))_{R^n} \tag{5.2}$$

where $(\quad|\quad)_{R^n}$ is the euclidean scalar product in R^n, $\nabla(foc^{-1})$ is the gradient in R^n, and $\alpha(c(x)) = (\alpha_1(c(x)), \ldots, \alpha_n(c(x)))$ is a vector in R^n. If $Z = \nabla_a f$, then

$$Z\phi = (\nabla_a\phi|Z)_a = (\nabla_a\phi|\nabla_a f)_a \tag{5.3}$$

In the chart c, let $a = a_{ij}$ be the expression of the Riemannian metric; the components of the vector field $\nabla_a f(x)$ are

$$\nabla_a f(x)_i = \sum_{j=1}^n (a^{-1}(c(x)))_{ij}\frac{\partial}{\partial c_j}(foc^{-1})(c(x)) \tag{5.4}$$

To verify (5.4), we consider $(\nabla_a f(x))g = \sum_{j=1}^n \beta_j(c(x))\frac{\partial}{\partial c_j}(goc^{-1})(c(x))$. We prove that $\beta_j(c(x))$ is given by (5.4). For $Z(x)h = \sum_{i=1}^n \alpha_i(c(x))\frac{\partial}{\partial c_i}(hoc^{-1})(c(x))$, we have by (3.1) that

$$(\nabla_a f(x)|Z(x))_a = \sum_{i,j} \beta_j(c(x))\alpha_i(c(x))a_{ij}(c(x)) = Z(x).f$$

By identification, we obtain $\beta_j(c(x))a_{ij}(c(x)) = \frac{\partial}{\partial c_i}(foc^{-1})(c(x))$. This yields (5.4).

From (5.4) and (3.1), we get

$$(\nabla_a f|\nabla_a\phi)_a = \sum_{i,j,k,p} a_{ij}(a^{-1})_{ip}\frac{\partial}{\partial c_p}(foc^{-1})(a^{-1})_{jk}\frac{\partial}{\partial c_k}(\phi oc^{-1})$$

$$= \sum_{j,k}(a^{-1})_{jk}\frac{\partial}{\partial c_j}(foc^{-1})\frac{\partial}{\partial c_k}(\phi oc^{-1}). \tag{5.5}$$

From (4.3), we obtain $div_{dv}(\nabla_a f(x))$ in local coordinates.

$$div_{dv}(\nabla_a f(x)) = \frac{1}{\sqrt{|\det a|}}\sum_{i,j}\frac{\partial}{\partial y_i}[\sqrt{|\det a|}(a^{-1})_{ij}\frac{\partial}{\partial y_j}(foc^{-1})(y)]_{|y=c(x)}. \tag{5.6}$$

We put $\Delta_a f(x) = \mathrm{div}_{dv}(\nabla_a f(x))$. The operator Δ_a is symmetric with respect to the Riemannian volume dv on M. *In the one dimensional case*, if $dv = \frac{dx}{\alpha(x)}$ is the Riemannian volume on R, $(ds^2 = \frac{dx^2}{\alpha(x)^2})$, then $a_{ij} = \frac{1}{\alpha^2}$ and $\Delta_a f(x) = \alpha\frac{\partial}{\partial y}(\alpha\frac{\partial}{\partial y}f)$.

6 Connections and divergences

We consider a connection on M and we denote by

$$\tau_t : T_{\phi(0)}(M) \to T_{\phi(t)}(M) \tag{6.1}$$

parallel transport on M along the differentiable curve $\phi(t)$. If $\phi(0) = x$ and $\frac{d}{dt}_{t=0}\phi(t) = Z(x)$, the covariant derivative ∇ associated to the connection is given by

$$\nabla_Z Y(x) = \frac{d}{dt}_{t=0} \tau_t^{-1} Y(\phi(t)). \tag{6.2}$$

The connection is Riemannian if for any vector fields Y and Z, and for any ϕ, it holds that

$$(\tau_t^{-1} Y(\phi(t)) | \tau_t^{-1} Z(\phi(t)))_{T_{\phi(0)}(M)} = (Y(\phi(t)) | Z(\phi(t)))_{T_{\phi(t)}(M)}. \tag{6.3}$$

For coordinate vector fields, we put

$$\nabla_{Z_i^c} Z_j^c = \gamma_{ij}^k Z_k^c. \tag{6.4}$$

The change of coordinates for the γ_{ij}^k is

$$\gamma_{\mu\nu}^{\lambda} \frac{\partial \tilde{c}_l}{\partial c_\lambda} = \tilde{\gamma}_{ij}^l \frac{\partial \tilde{c}_i}{\partial c_\mu} \cdot \frac{\partial \tilde{c}_j}{\partial c_\nu} + \frac{\partial^2 \tilde{c}_l}{\partial c_\mu \partial c_\nu}. \tag{6.5}$$

For two vector fields Y and Z, we put

$$T(Y, Z) = \nabla_Y Z - \nabla_Z Y - [Y, Z]. \tag{6.6}$$

From connection relations

$$\nabla_{fY} Z = f \nabla_Y Z \quad \text{and} \quad \nabla_Y (hZ) = Y(h) Z + h \nabla_Y Z, \tag{6.7}$$

it is well known that

$$T(fY, hZ) = fhT(Y, Z). \tag{6.8}$$

Lemma. *Consider a connection on M, not necessarily a Riemannian connection. Given the vector field Y, the trace of the covariant derivatives*

$$\mathcal{A}(Y) = \sum_i (\nabla_{Z_i^c} Y)^i \tag{6.9}$$

where$(\nabla_{Z_i^c} Y)^i$ is the component of $\nabla_{Z_i^c} Y$ along Z_i^c is independent of the chart c. Similarly, the traces

$$\mathcal{B}(Y) = \sum_k (\nabla_Y Z_k^c)^k + \sum_k [Z_k^c, Y] c_k \tag{6.10}$$

and

$$\mathcal{T}(Y) = \sum_i (T(Y, Z_i^c))^i = \mathcal{B}(Y) - \mathcal{A}(Y) \tag{6.11}$$

do not depend on the chart c.

Proof. Let $Y = \sum_{j=1}^n (\beta_j oc) Z_j^c$. From (6.7), we deduce

$$\nabla_{Z_i^c} Y = \sum_k [\sum_j (\beta_j oc)\gamma_{ij}^k + Z_i^c(\beta_k oc)] Z_k^c.$$

Thus

$$\mathcal{A}(Y) = \sum_i (\nabla_{Z_i^c} Y)^i = \sum_j (\beta_j oc)(\sum_i \gamma_{ij}^i) + \sum_i Z_i^c(\beta_i oc). \tag{6.12}$$

With the change of coordinates (6.5), we have

$$\sum_{ij} \widetilde{\beta}_j \widetilde{\gamma}_{ij}^i = \beta_k \frac{\partial \widetilde{c}_j}{\partial c_k}\left(\frac{\partial \widetilde{c}_i}{\partial c_l}\gamma_{\alpha\beta}^l \frac{\partial c_\alpha}{\partial \widetilde{c}_i}\frac{\partial c_\beta}{\partial \widetilde{c}_j} + \frac{\partial \widetilde{c}_i}{\partial c_l} \times \frac{\partial^2 c_l}{\partial \widetilde{c}_i \partial \widetilde{c}_j}\right).$$

This is equal to

$$\beta_k \gamma_{\alpha k}^\alpha + \beta_k \frac{\partial \widetilde{c}_j}{\partial c_k}\frac{\partial \widetilde{c}_i}{\partial c_l}\frac{\partial^2 c_l}{\partial \widetilde{c}_i \partial \widetilde{c}_j}. \tag{6.13}$$

From (6.13)–(1.3), we see that (6.9) does not depend on the choice of the chart c. With the same method, using (6.5)–(1.3), we prove that $\mathcal{B}(Y)$ is independent of c. This proves the lemma.

From now, we shall denote the connections with γ and the associated Christoffel symbols γ_{ij}^k. If the connection γ is Riemannian, and $\phi(t)$ is a curve on M such that $\frac{d}{dt}_{t=0}\phi(t) = Z_p^c(\phi(0))$, from (3.3) and then (6.2)–(6.3), we deduce that

$$\frac{d}{dt}_{t=0} a_{ik}(c(\phi(t))) = \frac{d}{dt}_{t=0}(Z_i^c(\phi(t))|Z_k^c(\phi(t)))$$

$$= \frac{d}{dt}_{t=0}(\tau_t^{-1} Z_i^c(\phi(t))|\tau_t^{-1} Z_k^c(\phi(t)))$$

$$= (\frac{d}{dt}_{t=0}\tau_t^{-1} Z_i^c(\phi(t))|\tau_t^{-1} Z_k^c(\phi(t))) + (\tau_t^{-1} Z_i^c(\phi(t))|\frac{d}{dt}_{t=0}\tau_t^{-1} Z_k^c(\phi(t)))$$

$$= (\nabla_{Z_p^c} Z_i^c | Z_k^c) + (Z_i^c | \nabla_{Z_p^c} Z_k^c) = \gamma_{pi}^j a_{jk} + \gamma_{pk}^j a_{ij}. \tag{6.14}$$

If γ is Riemannian, then

$$\frac{d}{dc_k} a_{ij} = \gamma_{ki}^l a_{lj} + \gamma_{kj}^l a_{il}, \tag{6.15}$$

Proposition. *Let γ be a connection on the Riemannian manifold M, and let dv be the Riemannian volume on M. Let a_{ij} be the expression in the chart c of the Riemannian metric, and a^{ij} the inverse matrix of a_{ij}. The connection γ satisfies the relation*

$$\mathcal{B}(Y) = div_{dv}(Y) \tag{6.16}$$

if and only if

$$\sum_i \gamma^i_{ki} = -\frac{1}{2}\sum_{i,j} a_{ij}\frac{\partial}{\partial c_k}a^{ij}. \tag{6.17}$$

Let γ be a Riemannian connection on the Riemannian manifold M; then it satisfies the trace condition (6.16)–(6.17). However, there are non-Riemannian connections on M for which (6.16)–(6.17) is true.

Proof. By definition of the Riemannian volume (4.2), we have

$$\frac{\partial}{\partial c_k}\log[\frac{d(c_*dv)}{dy}] = \frac{1}{2}\frac{\partial}{\partial c_k}\log|\det a_{ij}|.$$

With the formula (1.16) for the derivation of a determinant, we deduce that

$$\frac{\partial}{\partial c_k}\log|\det a_{ij}| = -\sum_{ij} a_{ij}\frac{\partial}{\partial c_k}a^{ij}.$$

With (4.3) this gives

$$div_{dv}(Z) = -\frac{1}{2}\sum_{i,j} a_{ij} Z(a^{ij} o c) + \sum_i [Z^c_i, Z] c_i. \tag{6.18}$$

On the other hand, if $Z = \sum_i (\alpha_i o c) Z^c_i$, then $\nabla_Z Z^c_k = \sum_i (\alpha_i o c)\gamma^j_{ik} Z^c_j$ and

$$\sum_k (\nabla_Z Z^c_k)^k = \sum_{k,i} (\alpha_i o c)\gamma^k_{ik}. \tag{6.19}$$

Comparing (6.18) and (6.19), we obtain the equivalence of (6.16) and (6.17).

If the connection γ is Riemannian, from (6.15) we have (6.17). For an example of a non-Riemannian connection for which (6.16)–(6.17) holds, see part 2, (7.25).

If γ is the Levi-Civita connection on M, and dv is the Riemannian volume on M, we have $\sum_i(\gamma^i_{ik} - \gamma^i_{ki}) = 0$; then, we use (6.17) to deduce that

$$\sum_i \gamma^i_{ik} = \frac{\partial}{\partial c_k}\log[\frac{d(c_*dv)}{dy}]. \tag{6.20}$$

We can also obtain (6.20) from Eisenhart[p.18, (7.9)] and from (4.2).

Corollary. *Let γ be a Riemannian connection on M and ∇ the covariant derivative associated to γ. Let dv be the Riemannian volume on M. For a vector field Y, we have*

$$div_{dv}(Y) = \sum_i (\nabla_{Z^c_i} Y)^i + \mathcal{T}(Y) = \mathcal{A}(Y) + \mathcal{T}(Y). \tag{6.21}$$

If T^i_{kj} is the torsion associated to a Riemannian connnection on M, we have, for a vector field $Y = \sum_j (\beta_j o c) Z^c_j$,

$$div_\mu(Y) = Y(\log \frac{d\mu}{dv}) + \sum_j (\beta_j oc)(\sum_k T^k_{kj}) + \sum_i (\nabla_{Z^c_i} Y)^i, \tag{6.22}$$

where μ is a measure absolutely continuous with respect to the Riemannian volume dv. Moreover, if γ is the Levi-Civita connection, then

$$div_\mu(Y) = Y(\log \frac{d\mu}{dv}) + \sum_i (\nabla_{Z^c_i} Y)^i. \tag{6.23}$$

The identity (6.23) remains true for any Riemannian connection γ, satisfying the torsion condition $\mathcal{T} = 0$ i.e., $\sum_i (\gamma^i_{ij} - \gamma^i_{ji}) = \sum_i T^i_{ij} = 0$.

Remark. $\mathcal{T}(Y)$ does not depend on the metric on M; the condition $\mathcal{T}(Y) = 0$ is different from the Driver's condition since the Driver's condition is related to the Riemannian metric on M. See [2].

Part II Direct image operators

Let M and $\widetilde{M}$ be two differentiable manifolds. We assume that the manifold M is n-dimensional and $\widetilde{M}$ is p-dimensional, with $p \leq n$. Let $g : M \to \widetilde{M}$ be a differentiable map such that for any $m \in M$ the differential $g'(m)$ is surjective. We give a probability measure μ on M. The image measure on $\widetilde{M}$ is $g_*\mu$.

1 Image of functions

The direct image is constructed by using conditional expectation. See [5 p.76–82]. If $f : M \to R$, the density of the measure $g_* f\mu$ with respect to the measure $g_*\mu$ is denoted by

$$E^g_\mu(f) = \frac{d(g_* f\mu)}{d(g_*\mu).}$$

Let $f : M \to R$, we define the direct image $g_*(f) : \widetilde{M} \to R$ with

$$g_*(f)(\xi) = E^g_\mu(f)(\xi). \tag{1.1}$$

For $\widetilde{f} : \widetilde{M} \to R$, we define the inverse image $g^*(\widetilde{f})$ as

$$g^*(\widetilde{f}) = \widetilde{f}og. \tag{1.2}$$

Since $E^g_\mu(\widetilde{f}og) = \widetilde{f}$, we have $g_*g^* = Id_{\text{functions on } \widetilde{M}}$. To calculate g^*g_*, we give an example: Let $M = R^2$ with probability measure $\mu = \exp(-\frac{x^2+y^2}{2})\frac{dxdy}{2\pi}$.

Let $\widetilde{M} = R$ with the measure $\widetilde{\mu} = \exp(-\frac{z^2}{2})\frac{dz}{\sqrt{2\pi}}$. We define $g : M \to \widetilde{M}$ by $g(x, y) = x$. Then $g_*\mu = \widetilde{\mu}$. For $f : R^2 \to R$, it is easy to compute that

$$g_*(f)(x) = \int f(x, y)\exp(-\frac{y^2}{2})\frac{dy}{\sqrt{2\pi}};$$

thus, in general, $g^*g_* \neq Id_{\text{functions on } M}$. However for $f : M \to R$, we have

$$\int g^*g_*(f)d\mu = \int f d\mu.$$

2 Image of vector fields. Divergences

Let Z be a vector field on M. We may denote either $Zf(x)$ or $Z(x)f$. For a differentiable function $\widetilde{f} : \widetilde{M} \to R$, we have

$$< g'(x)Z(x), d\widetilde{f}(x) >= Z(\widetilde{f}og)(x). \tag{2.1}$$

We define the vector field g_*Z on $\widetilde{M}$ with

$$g_*(Z)\widetilde{f}(\xi) = E^g_\mu[Z(\widetilde{f}og)](\xi). \tag{2.2}$$

With (0.1), we can also write (2.2) as

$$E^g_\mu[Z(\widetilde{f}og)] = g_*(Z(g^*\widetilde{f})). \tag{2.3}$$

Let $\widetilde{f}$ and $\widetilde{\phi}$ be two differentiable functions from $\widetilde{M}$ to R. We have

$$(g_*Z)(\widetilde{f\phi}) = ((g_*Z)\widetilde{f})\widetilde{\phi} + \widetilde{f}((g_*Z)\widetilde{\phi}) \tag{2.4}$$

Thus (2.2) defines a vector field. The identity $E^g_\mu[(\widetilde{f}og)Z] = \widetilde{f}E^g_\mu(Z)$ can be written as (0.18). In local coordinates, if $Zf(x) = \sum_{i=1}^n \alpha_i(c(x))\frac{\partial}{\partial c_i}(foc^{-1})(c(x))$, and if $\widetilde{c}$ is a chart for $\widetilde{M}$, we have

$$(g_*Z)\widetilde{f}(\xi) = \sum_{j=1}^p \beta_j(\widetilde{c}(\xi))\frac{\partial}{\partial \widetilde{c}_j}(\widetilde{f}o\widetilde{c}^{-1})(\widetilde{c}(\xi))$$

with

$$\beta_j(\widetilde{c}(\xi)) = \sum_{i=1}^n E^g_\mu[\alpha_i(c(x))\frac{\partial(\widetilde{c}ogoc^{-1})_j}{\partial c_i}(c(x))](\xi). \tag{2.5}$$

(2.5) can be written as $g_*Z = \sum_j g_*(Z(g^*\widetilde{c}_j))Z^{\widetilde{c}}_j$. To prove (2.5), we write

$$(g_*Z)\widetilde{f}(\xi) = E^g_\mu[Z(\widetilde{f}og)(x)] = E^g_\mu[\sum_{i=1}^n \alpha_i(c(x))\frac{\partial}{\partial c_i}(\widetilde{f}ogoc^{-1})(c(x))](\xi).$$

Since

$$\frac{\partial}{\partial c_i}(\tilde{f}ogoc^{-1})(c(x)) = \frac{\partial}{\partial c_i}(\tilde{f}o\tilde{c}^{-1}o\tilde{c}ogoc^{-1})(c(x))$$

$$= \sum_j \frac{\partial}{\partial \tilde{c}_j}(\tilde{f}o\tilde{c}^{-1})(\tilde{c}(g(x))\frac{\partial(\tilde{c}_j ogoc^{-1})}{\partial c_i}(c(x))$$

we obtain (2.5). Let $\tilde{d}$ be another chart and $\delta_i(\tilde{d}(\xi))$ the coordinates in this new chart; we have

$$\beta_j(\tilde{c}(\xi)) = \sum_i \delta_i(\tilde{d}(\xi))[\frac{\partial(\tilde{c}o\tilde{d}^{-1})_j}{\partial \tilde{d}_i}](\tilde{d}(\xi));$$

thus $\beta_j(\tilde{c}(\xi))$ satisfies the change of coordinates formula for vector fields (1.3).

Proposition. *Let Z be a vector field on M. Then (0.10)*

$$div_{g_*\mu}(g_*Z) = g_*(div_\mu Z). \tag{2.6}$$

Proof. We write (2.6) as $div_{g_*\mu}[E^g_\mu(Z)] = E^g_\mu[div_\mu(Z)]$. The identity is a consequence of part 1 (2.1).

Remark. Let $\tilde{Z}$ be a vector field on $\tilde{M}$; to put $(Zf)(x) = \tilde{Z}(g_*f)(g(x))$ does not define a vector field on M since $Z(f\phi) \neq fZ\phi + \phi Zf$.

3 Inverse image of vector fields

In the following, we define the inverse image of a vector field using the Riemannian metric on M. For that, we assume that the jacobian map $g'(x) : T_x(M) \to T_{g(x)}(\tilde{M})$ is *surjective*; the restriction of $g'(x)$ to the orthogonal of $\ker(g'(x))$ is an isomorphism. Let $(g'(x))^{-1}$ be its inverse. See [5 p. 70] for the concept of lifting up and pushing down vector fields.

Let $\tilde{Z}$ be a vector field on $\tilde{M}$. We define the vector field $Z = g^*\tilde{Z}$ on M with

$$Z(x) = g'(x)^{-1}(\tilde{Z}(g(x))) \tag{3.1}$$

and $Z(x)$ is in the orthogonal of the subspace $Ker g'(x)$ of $T_x(M)$ endowed with the Riemannian scalar product.

From (3.1), we deduce (see (0.3) and $(0.18)_1$ that

$$g^*(\tilde{Z})(g^*\tilde{f}) = g^*(\tilde{Z}\tilde{f}) \quad \text{and} \quad g^*(\tilde{f}\tilde{Y}) = g^*(\tilde{f})g^*(\tilde{Y}). \tag{3.2}$$

Let $Z = g^*(\tilde{Z})$; then by (3.1)–(3.2)–(2.1)

$$g_*g^*(\tilde{Z})\tilde{f} = E^g_\mu[g^*(\tilde{Z})(\tilde{f}og)] = E^g_\mu[Z(x)(\tilde{f}og)]$$

We deduce that $g_* g^* = Id_{\text{vector fields on} \widetilde{M}}$. See (0.7). It can be seen from the following example that $g^* g_* \neq Id_{\text{vector fields on} M}$.

Example (3.3). Let $M = R^2$ with the measure $\mu = \exp(-\frac{x^2+y^2}{2})\frac{dxdy}{2\pi}$. Let $\widetilde{M} = R$ and $g(x, y) = x^2 + y^2$. Then $g_*\mu = 1_{[0,+\infty[}e^{-u/2}\frac{du}{4\pi}$. We see that $E^g_\mu(x) = 0$ since x is an odd function, but $E^g_\mu(x^2) \neq 0$. From this we deduce that $g_*(\frac{\partial}{\partial x}) = 0$, but $g_*(x\frac{\partial}{\partial x}) \neq 0$. Remark that we also may have $g_* Z = 0$ and $g_*(\phi Z) \neq 0$.

From (2.1) Part I, we have

$$\widetilde{Z}(g_* f) = div_{g_*\mu}((g_* f)\widetilde{Z}) - (g_* f) \quad div_{g_*\mu}(\widetilde{Z}).$$

This identity, with (0.3) and (0.10) implies $(0.11)_2$. We can also verify $(0.11)_2$ in the example 7 below. See (7.10)–(7.11). Let $\phi(x, y, \theta) = y$; we have $g_*\phi(x, u) = (1 + \epsilon^2)^{-1}u$ and also $\frac{\partial}{\partial u} g_*\phi = (1 + \epsilon^2)^{-1}$. See (7.14). On the other hand, from (7.18) we see that

$$g^*(\frac{\partial}{\partial u})(\phi) = (1 + \epsilon x)(\epsilon^2 + (1 + \epsilon x)^2)^{-1}.$$

In the following, we calculate in local coordinates, the inverse image of a vector field. Let $\widetilde{c}$ be a chart on $\widetilde{M}$. Let a be the metric on M. For a function $f : M \to R$, we denote $\nabla_a f$ the gradient of f with the metric a. We put

$$\lambda_{ip} = (\nabla_a(\widetilde{c}_i o g)|\nabla(\widetilde{c}_p o g))_a. \tag{3.4}$$

We denote by λ the matrix (λ_{ip}). The coordinate vector field $Z^{\widetilde{c}}_p$ in the chart $\widetilde{c}$ is

$$Z^{\widetilde{c}}_p \widetilde{f} = \frac{\partial}{\partial \widetilde{c}_p}(\widetilde{f} o \widetilde{c}^{-1}). \tag{3.5}$$

Lemma.

$$g^*(Z^{\widetilde{c}}_p) = \sum_i \lambda^{-1}_{ip} \nabla_a(\widetilde{c}_i o g) \tag{3.6}$$

Moreover

$$(g^*(Z^{\widetilde{c}}_i)|g^*(Z^{\widetilde{c}}_j))_a = (\lambda^{-1})_{ij} \tag{3.7}$$

Proof. We put $Z = \sum_i \lambda^{-1}_{ip} \nabla_a(\widetilde{c}_i o g)$. We verify first that $Z(x)$ is in the orthogonal of $\ker g'(x)$. Let $Z_1(x)$ be such that $g'(x)Z_1(x) = 0$; then by (5.5) Part I and since

$$Z_1(x)(\widetilde{c}_i o g) = [g'(x)Z_1(x)](\widetilde{c}_i) = 0$$

we have

$$(Z(x)|Z_1(x))_a = \sum_i \lambda^{-1}_{ip}(\nabla_a(\widetilde{c}_i o g)|Z_1(x))_a = \sum_i \lambda^{-1}_{ip} Z_1(x)(\widetilde{c}_i o g) = 0.$$

Now, we verify that $g'(x)Z(x) = Z_p^{\widetilde{c}}(g(x))$.

$$g'(x)Z(x)\widetilde{f} = Z(x)(\widetilde{f}og) = \sum_i \lambda_{ip}^{-1}(\nabla_a(\widetilde{c}_i og)|\nabla_a(\widetilde{f}og))_a.$$

By (5.5) Part I, this gives

$$g'(x)Z(x)\widetilde{f} = \sum (a^{-1})_{jk}\lambda_{ip}^{-1}\frac{\partial}{\partial c_j}(\widetilde{c}_i ogoc^{-1})\frac{\partial}{\partial c_k}(\widetilde{c}_q ogoc^{-1})\frac{\partial}{\partial \widetilde{c}_q}(\widetilde{f}o\widetilde{c}^{-1})$$

$$= \sum \lambda_{ip}^{-1}\lambda_{iq}\frac{\partial}{\partial \widetilde{c}_q}(\widetilde{f}o\widetilde{c}^{-1}) = \frac{\partial}{\partial \widetilde{c}_p}(\widetilde{f}o\widetilde{c}^{-1}).$$

This proves (3.6). As an example, consider $g : R^2 \to R$ where $g(x, y) = x^2 + y^2$. From $g'(x, y) = (2x, 2y)$, the orthogonal of $\ker g'(x_o, y_o)$ is the set of vectors colinear to (x_o, y_o). Let

$$\widetilde{Z} = \alpha(\xi)\frac{d}{d\xi} \quad \text{then} \quad g^*(\widetilde{Z}) = \frac{\alpha(x^2+y^2)}{2(x^2+y^2)}(x\frac{d}{dx} + y\frac{d}{dy}). \tag{3.8}$$

Proposition. *The property (1.19)-Part I is stable by g^*. Let $(\widetilde{v}_i)_{i=1,\dots,k}$ be a set of independent vector fields on $\widetilde{M}$ and assume that $\widetilde{w}_j = \sum_{i=1}^k b^i_j \widetilde{v}_i$ for $j = 1, \dots, k$ with $\det b^i_j \neq 0$. We assume also that for any i, j,*

$$[\widetilde{w}_i, \widetilde{w}_j] = \sum_{s,p} b_i^s b_j^p [\widetilde{v}_s, \widetilde{v}_p]. \tag{3.9}$$

Then

$$g^*(\widetilde{w}_j) = \sum_i (b^i_j og) g^*(\widetilde{v}_i) \tag{3.10}$$

$$[g^*(\widetilde{w}_i), g^*(\widetilde{w}_j)] = \sum_{s,p} (b_i^s og)(b_j^p og)[g^*(\widetilde{v}_s), g^*(\widetilde{v}_p)]. \tag{3.11}$$

Proof. (3.10) comes from (3.2). From (3.9), we deduce that

$$[g^*(\widetilde{w}_p), g^*(\widetilde{w}_j)] = \sum_i \left(g^*(\widetilde{w}_p)(b^i_j og) - g^*(\widetilde{w}_j)(b^i_p og)\right)g^*(\widetilde{v}_i)$$

$$+ \sum_{k,s} (b_i^k og)(b_j^s og)[g^*(\widetilde{v}_k), g^*(\widetilde{v}_s)].$$

Since $\widetilde{w}_p(b^i_j) = \widetilde{w}_j(b^i_p)$ and due to (3.2), we get $g^*(\widetilde{w}_p)(b^i_j og) = g^*(\widetilde{w}_j)(b^i_p og)$. This proves (3.11).

In the following, we keep in mind the case where $(\widetilde{w}_i)_{(i=1,\dots,p)}$ is the system of coordinates vector fields in a chart of $\widetilde{M}$. See (1.18)–(1.19) Part I.

Let $(\widetilde{w}_i)_{(i=1,\dots,k)}$ be a system of k independent vector fields on $\widetilde{M}$, and let $\widetilde{\mathcal{W}}$ be the class of vector fields equivalent to $(\widetilde{w}_i)_{(i=1,\dots,p)}$ as in (1.18)–(1.19) Part I. We assume that $\widetilde{\psi}$ is a tensor of type (0,2) on $\widetilde{M}$. We denote $\widetilde{\psi}_x$ to be the associated bilinear form on $T_x(\widetilde{M}) \times T_x(\widetilde{M})$; we assume $\det \widetilde{\psi} \neq 0$. See (1.25) Part I. For $\widetilde{Z} = \sum_i \alpha_i \widetilde{w}_i$, we put as in Part I(1.26)

$$Trace_{\widetilde{Z}}^{\widetilde{\psi}}(\widetilde{\mathcal{W}}) = \sum_i \widetilde{w}_i(\alpha_i) + \frac{1}{2}Z(\log|\det_{\widetilde{w}_i}(\widetilde{\psi})|). \tag{3.12}$$

Lemma. *We assume that there exists a tensor of type (0,2) on M, i.e., a bilinear form ψ, such that for any vector field $\widetilde{Z}$ on $\widetilde{M}$, we have*

$$\widetilde{Z}(\log|\det_{\widetilde{w}_i} \widetilde{\psi}|) = g_*[g^*(\widetilde{Z})(\log|\det_{g^*(\widetilde{w}_i)} \psi)|]. \tag{3.13}$$

Then (3.13) remains true if we replace $(\widetilde{w}_i)_{(i=1,\dots,k)}$ by any equivalent system of vectors.

Proof. Assume that $(\widetilde{w}_i)_{(i=1,\dots,k)}$ and $(\widetilde{v}_i)_{(i=1,\dots,k)}$ are in the same class of equivalence. See (1.18)–(1.19) Part 1. We have $\widetilde{w}_j = \sum_i b^i_j \widetilde{v}_i$ and

$$g^*(\widetilde{w}_j) = \sum_i (b^i_j o g) g^*(\widetilde{v}_i).$$

We use the property that ψ and $\widetilde{\psi}$ are bilinear on each fiber of the tangent bundle; we put $\widetilde{f} = \log|\det b^j_i|$; then by (0.7), we have $\widetilde{Z}\widetilde{f} = g_*(g^*(\widetilde{Z})(\widetilde{f} o g))$. This proves (3.13) for the $(\widetilde{v}_i)$.

Notation. Let $\widetilde{\mathcal{W}}$ be an equivalence class of vector fields on $\widetilde{M}$. We denote $g^*(\widetilde{\mathcal{W}})$ as the set of system of vector fields of the form $(g^*(\widetilde{w}_i)_{i=1,\dots,k}$ where $(\widetilde{w}_i))_{i=1,\dots,k} \in \widetilde{\mathcal{W}}$. By (3.10)–(3.11), the systems of vectors in $g^*(\widetilde{\mathcal{W}})$ are equivalent.

Let $(g^*(\widetilde{w}_i))_{i=1,\dots,k}) \in g^*(\widetilde{\mathcal{W}})$; for a vector field $Z = \sum_j \alpha_j g^*(\widetilde{w}_j)$, we put

$$Trace_Z^{\psi}(g^*(\widetilde{\mathcal{W}})) = \sum_i g^*(\widetilde{w}_i)(\alpha_i) + \frac{1}{2}Z(\log|\det_{g^*(\widetilde{w}_i)} \psi|) \tag{3.14}$$

and we assume that (3.13) is true. We conjecture that

$$Trace_{\widetilde{Z}}^{\widetilde{\psi}}(\widetilde{\mathcal{W}}) = g_*(Trace_{g^*(\widetilde{Z})}(g^*(\widetilde{\mathcal{W}}))). \tag{3.15}$$

4 Image of a metric

In [1], the image of a metric was defined. In the following, the method to construct the image metric $b = g_* a$ is different from that of [1]. With assumptions on a, we show that the image metric $b = g_*a$ is the same as the one obtained in [1] if we restrict [1] to the finite dimensional case.

Definition. For two vector fields $\widetilde{Y}$ and $\widetilde{Z}$ on $\widetilde{M}$, we put (0.15)

$$(\widetilde{Y}|\widetilde{Z})_{g_*a} = g_*((g^*(\widetilde{Y})|g^*(\widetilde{Z}))_a) \tag{4.1}$$

For a function $\widetilde{f} : \widetilde{M} \to R$, we verify that $(\widetilde{f}\widetilde{Y}|\widetilde{Z})_{g_*a} = \widetilde{f}(\widetilde{Y}|\widetilde{Z})_{g_*a}$. Thus, we see that (4.1) is bilinear and defines a scalar product. We denote by $a = (a_{ij})$ the metric on M and ∇_a the associated gradient. See (5.1)-(5.3)-(5.4) part1. Let $(a^{ij}) = (a_{kp}^{-1})$ be the inverse matrix of $a = a_{kp}$; then

$$(\nabla_a f(x)|\nabla_a \phi(x))_{T_x(M)} = a^{ij}(c(x))\frac{\partial}{\partial c_i}(foc^{-1})(c(x))\frac{\partial}{\partial c_j}(\phi oc^{-1})(c(x)) \tag{4.2}$$

Lemma. *Let $\widetilde{c}$ be a chart on $\widetilde{M}$. Consider the matrix $\lambda_{pq} = (\nabla_a(\widetilde{c}_p og)|\nabla_a(\widetilde{c}_q og))_a$. See (3.4). Let $b = g_*a$ be the metric defined by (4.1). Then, in the chart $\widetilde{c}$,*

$$b_{ij} = (Z_i^{\widetilde{c}}|Z_j^{\widetilde{c}})_b = E_\mu^g((\lambda^{-1})_{ij}) \tag{4.3}$$

Proof. By (4.1), we have $(Z_i^{\widetilde{c}}|Z_j^{\widetilde{c}})_b = E_\mu^g((g^*(Z_i^{\widetilde{c}})|g^*(Z_j^{\widetilde{c}}))_a)$. Then we use (3.6).

Proposition. *Let $b = g_*a$ be a image metric on $\widetilde{M}$. For the gradients of functions, we have (see (0.14))*

$$\nabla_{g_*a}\widetilde{\phi} = g_*(\nabla_a(g^*\widetilde{\phi})) \tag{4.4}$$

if and only if the expression of the image metric b in a chart $\widetilde{c}$ is given by

$$(b^{-1})_{kp} = E_\mu^g[(a^{-1})_{ij}(c(x))\frac{\partial(\widetilde{c}_p ogoc^{-1})}{\partial c_i}(c(x))\frac{\partial(\widetilde{c}_k ogoc^{-1})}{\partial c_j}(c(x))]$$

$$= E_\mu^g[(\nabla_a(\widetilde{c}_p og)|\nabla_a(\widetilde{c}_k og))_a](\xi) = E_\mu^g(\lambda_{kp}) \tag{4.5}$$

Proof. We can write (4.4) as

$$(\nabla_b\widetilde{\phi})(\xi)\widetilde{f} = E_\mu^g[[\nabla_a(\widetilde{\phi}og)](\widetilde{f}og)](\xi) \tag{4.6}$$

We have (see (5.1)-(5.4) part1)

$$(\nabla_b\widetilde{\phi})\widetilde{f} = (b^{-1})_{pq}(\widetilde{c}(\xi))\frac{\partial}{\partial\widetilde{c}_p}(\widetilde{f}o\widetilde{c}^{-1})(\widetilde{c}(\xi))\frac{\partial}{\partial\widetilde{c}_q}(\widetilde{\phi}o\widetilde{c}^{-1})(\widetilde{c}(\xi)) \tag{4.7}$$

On the other hand,

$$\nabla_a(\widetilde{\phi}og)(\widetilde{f}og) = (a^{-1})_{ij}(c(x))\frac{\partial(\widetilde{f}ogoc^{-1})}{\partial c_i}(c(x))\frac{\partial(\widetilde{\phi}ogoc^{-1})}{\partial c_j}(c(x))$$

$$= (a^{-1})_{ij}(c(x))\frac{\partial(\widetilde{f}o\widetilde{c}^{-1})}{\partial\widetilde{c}_p}(\widetilde{c}(g(x))\frac{\partial(\widetilde{c}_p ogoc^{-1})}{\partial c_i}(c(x))$$

$$\times\frac{\partial(\widetilde{\phi}o\widetilde{c}^{-1})}{\partial\widetilde{c}_k}(\widetilde{c}(g(x))\frac{\partial(\widetilde{c}_k ogoc^{-1})}{\partial c_j}(c(x)). \tag{4.8}$$

Thus

$$E^g_\mu[(\nabla_a(\widetilde{\phi}og))(\widetilde{f}og)](\xi) = (b^{-1})_{kp}(\widetilde{c}(\xi))\frac{\partial(\widetilde{\phi}o\widetilde{c}^{-1})}{\partial\widetilde{c}_k}(\widetilde{c}(\xi))\frac{\partial(\widetilde{f}o\widetilde{c}^{-1})}{\partial\widetilde{c}_p}(\widetilde{c}(\xi)) \tag{4.9}$$

where $(b^{-1})_{kp}$ is given by (4.5). Then, we identify (4.7) and (4.9).

In [1], the image metric is defined with (4.5). We have to identify the metric b of (4.4)–(4.5) with the one defined by (4.1).

Lemma. *Consider the matrix $\lambda_{pq} = (\nabla_a(\widetilde{c}_p og)|\nabla_a(\widetilde{c}_q og))_a$. The metric defined by (4.1) satisfies (4.5) if and only if the inverse of the matrix $E^g_\mu(\lambda_{pk})$ is the matrix $E^g_\mu((\lambda^{-1})_{kj})$.*

Proof. We compare (4.3) and (4.5).

Proposition. *For two vector fields Y and Z on M, in general*

$$g_*((Y|Z)_a) \neq (g_*(Y)|g_*(Z))_{g_*a} \tag{4.10}$$

but for the vector fields $Y = \nabla_a(\widetilde{\phi}og)$ and $Z = \nabla_a(\widetilde{\psi}og)$ the equality is true in (4.10). See (0.15) and (0.16).

Proof. Let $Y = \nabla_a(\widetilde{\phi}og)$ and $Z = \nabla_a(\widetilde{\psi}og)$. By (4.4), we have $g_*(Y) = \nabla_a\widetilde{\phi}$ and $g_*(Z) = \nabla_a\widetilde{\phi}$. Since $(\nabla_a(\widetilde{\phi}og)|\nabla_a(\widetilde{\psi}og))_{g_*a} = [\nabla_a(\widetilde{\phi}og)](\widetilde{\psi}og)$, we obtain

$$E^g_\mu[[\nabla_a(\widetilde{\phi}og)](\widetilde{\psi}og)] = g_*(Y)\widetilde{\psi} = (\nabla_b\widetilde{\phi})\widetilde{\psi} = (\nabla_b\widetilde{\phi}|\nabla_b\widetilde{\psi})_b = (g_*(Y)|g_*(Z))_b$$

This proves the equality in (4.10) in the special case of the gradient vector fields Y and Z. We can see that the equality in (4.10)-(0.16) is not true in general with the example (3.3); we take $Z = x\frac{\partial}{\partial x} - y\frac{\partial}{\partial y}$ and $Y = x\frac{\partial}{\partial x}$. Then $g_*(Z) = 0$ but $g_*((Y|Z)_a) \neq 0$. For the example (3.3)–(3.8), on R^2, we take a to be the euclidean metric. Let $\widetilde{Y}(\xi) = \alpha(\xi)\frac{d}{d\xi}$ and $\widetilde{Z}(\xi) = \beta(\xi)\frac{d}{d\xi}$, we have

$$(\widetilde{Y}|\widetilde{Z})_{g_*a} = \frac{\alpha(\xi)\beta(\xi)}{4\xi} \tag{4.11}$$

We can prove (4.11) directly from (4.1) or we can use (4.5).

5 Image of a connection

Lemma. *Let $\widetilde{Y}$ be a vector field on $\widetilde{M}$, and $\widetilde{f}$ be a function from $\widetilde{M}$ to R. Then*

$$g^*(\widetilde{f}\widetilde{Y}) = g^*(\widetilde{f})g^*(\widetilde{Y}) \tag{5.1}$$

$$g_*(g^*(\tilde{f})Y) = \tilde{f}g_*(Y). \tag{5.2}$$

See $(0.18)_1 - -(0.18)_2$.

Proof. The vector field $Y = g^*(\tilde{Y})$ satisfies $g'(x)Y(x) = \tilde{Y}(g(x))$ and $Y(x)$ is in the orthogonal of $\ker g'(x)$. From (2.1), we have $[g'(x)Y(x)]\tilde{\phi} = Y(\tilde{\phi}og)(x)$. Since $g'(x)$ is linear

$$g'(x)((\tilde{f}og)(x)Y(x))\tilde{\phi} = \tilde{f}og(x)(\tilde{Y}\tilde{\phi})(g(x)) = (\tilde{f}\tilde{Y})(\tilde{\phi})(g(x))$$

This proves (5.1). The identity (5.2) is straightforward.

Lemma. *Let ∇ be the covariant derivative associated to a connection on M. For two vector fields $\tilde{Y}$ and $\tilde{Z}$ on $\tilde{M}$, we put*

$$\tilde{\nabla}_{\tilde{Y}}\tilde{Z} = g_*(\nabla_{g^*(\tilde{Y})}(g^*(\tilde{Z}))). \tag{5.3}$$

Then (5.3) defines the covariant derivative associated to a connection on $\tilde{M}$.

Proof. Let $\tilde{f} : \tilde{M} \to R$. From (5.2) and (5.3) and then (5.1), we have

$$\tilde{\nabla}_{\tilde{f}\tilde{Y}}\tilde{Z} = g_*(\nabla_{(\tilde{f}og)g^*(\tilde{Y})}(g^*(\tilde{Z}))) = g_*(\tilde{f}og\nabla_{g^*(\tilde{Y})}(g^*(\tilde{Z}))) = \tilde{f}\tilde{\nabla}_{\tilde{Y}}\tilde{Z}. \tag{5.4}$$

$$\tilde{\nabla}_{\tilde{Y}}(\tilde{f}\tilde{Z}) = g_*(\nabla_{g^*(\tilde{Y})}(g^*(\tilde{f}\tilde{Z}))) = g_*(\nabla_{g^*(\tilde{Y})}((\tilde{f}og)g^*(\tilde{Z}))). \tag{5.5}$$

Since

$$\nabla_X(fY) = f\nabla_X Y + (Xf)Y$$

we deduce from (5.5) and (3.2) that

$$\begin{aligned}\tilde{\nabla}_{\tilde{Y}}(\tilde{f}\tilde{Z}) &= g_*((\tilde{f}og)\nabla_{g^*(\tilde{Y})}(g^*(\tilde{Z})) + \big(g^*(\tilde{Y})(\tilde{f}og)\big)g^*(\tilde{Z}) \\ &= \tilde{f}\tilde{\nabla}_{\tilde{Y}}\tilde{Z} + (\tilde{Y}\tilde{f})\tilde{Z}.\end{aligned}$$

Lemma. *When $g : R^2 \to R$ is given by $g(x, y) = x^2 + y^2$, any Riemannian connection ∇ on R^2 with the euclidean metric has for image a connection $g_*\nabla$ which is Riemannian for the image metric.*

Proof. See (3.3)–(3.8)–(4.11) where R^2 is endowed with the euclidean metric. On R^2, we take any Riemannian connection ∇. The Christoffel symbols γ_{ij}^k satisfy (6.15) Part I. In that case, we show that the image connection $\tilde{\nabla}$ is Riemannian on R with the image metric (4.1). We put (see (3.8))

$$Z = \frac{1}{2(x^2+y^2)}(x\frac{\partial}{\partial x} + y\frac{\partial}{\partial y}) = g^*(\frac{d}{d\xi}) \tag{5.6}$$

then

$$\nabla_Z Z = -\frac{1}{4(x^2+y^2)^2}(x\frac{\partial}{\partial x} + y\frac{\partial}{\partial y}) + \frac{x\gamma_{12}^1 + y\gamma_{22}^1}{4(x^2+y^2)^2}(y\frac{\partial}{\partial x} - x\frac{\partial}{\partial y}). \tag{5.7}$$

With (3.3) and since $E^g_\mu(\phi(x,y)(y\frac{\partial}{\partial x} - x\frac{\partial}{\partial y})) = 0$, we see that

$$\widetilde{\nabla}_{\frac{d}{d\xi}} \frac{d}{d\xi} = g_*(\nabla_Z Z) = -\frac{1}{2\xi}\frac{d}{d\xi}. \tag{5.8}$$

The Christoffel symbol for $\widetilde{\nabla}$ is given by $\widetilde{\gamma}^1_{11} = -\frac{1}{2\xi}$.

From (6.15) Part I, the connection $\widetilde{\nabla}$ is Riemannian if $\frac{d}{d\xi}g_{11} = \gamma^1_{11}g_{11} + \gamma^1_{11}g_{11}$. This last identity is satisfied since by (4.1), the image metric is $g_{11} = \frac{1}{4\xi}$. We have $\frac{d}{d\xi}(\frac{1}{4\xi}) = 2(-\frac{1}{2\xi}) \times \frac{1}{4\xi}$. This proves the lemma.

Remark. Let $\widetilde{Y} = \widetilde{U} = \frac{d}{d\xi}$, we verify on the example (3.3)–(3.8)–(4.11) that

$$g^*(\widetilde{\nabla}_{\widetilde{U}}\widetilde{Y}) = \nabla_{g^*(\widetilde{U})}g^*(\widetilde{Y}), \tag{5.9}$$

where $W \in \ker g'(x)$. For this example, we have $W = 0$.

Proposition. *Let $\widetilde{Y}$, $\widetilde{Z}$, and $\widetilde{U}$ be vector fields on $\widetilde{M}$. Assume that for $\widetilde{U}$ and $\widetilde{Y}$, we have $g^*(\widetilde{\nabla}_{\widetilde{U}}\widetilde{Y}) = \nabla_{g^*(\widetilde{U})}g^*(\widetilde{Y}) + W$, where $W \in \ker g'(x)$. Compare with (5.9); then for the covariant derivative $\widetilde{\nabla}$ of the image connection, we have*

$$(\widetilde{\nabla}_{\widetilde{U}}\widetilde{Y}|\widetilde{Z})_{g_*a} = g_*((\nabla_{g^*(\widetilde{U})}g^*(\widetilde{Y})|g^*(\widetilde{Z}))_a) \tag{5.10}$$

Proof. (5.10) is a consequence of (5.9) and (3.1) since $g^*(\widetilde{U})$ is in the orthogonal of $\ker g'(x)$. From(0.7), we have $\widetilde{U} = g_*g^*(\widetilde{U})$. We put $Z = g^*(\widetilde{Z})$ and $Y = g^*(\widetilde{Y})$; we obtain

$$(\widetilde{\nabla}_{\widetilde{Y}}\widetilde{Z}|\widetilde{U}) = (g_*(\nabla_Y Z)|g_*g^*(\widetilde{U})) = E^g_\mu[(\nabla_Y Z|g^*(\widetilde{U}))].$$

In the following, we define the image of a curve C on M, the inverse image of a curve $\widetilde{C}$ on $\widetilde{M}$, the image of a parallel transport of a vector field along a curve, and we see why the image of a Riemannian connection is not always Riemannian with the image metric.

Definition. (Direct image of a curve). Let $C : t \to \phi(t)$ be a curve on M, we denote g_*C the curve on $\widetilde{M}$ given by $g_*C : t \to g(\phi(t))$.

Definition. (Inverse image of a curve). Let $\widetilde{C}_{g(x_0)} : t \to \widetilde{\phi}(t)$ be a curve on $\widetilde{M}$ such that $\widetilde{\phi}(0) = g(x_0)$. We define a curve $C_{x_0} : t \to \phi(t)$ on M by solving the differential equation

$$g'(\phi(t))\frac{d}{dt}\phi(t) = \frac{d}{dt}\widetilde{\phi}(t) \tag{5.11}$$

with the condition that $\frac{d}{dt}\phi(t)$ be in the orthogonal of $\ker(g'(\phi(t)))$ and the initial data $\phi(0) = x_0$.

Lemma.

$$g_*g^*(\widetilde{C}_{g(x_0)}) = \widetilde{C}_{g(x_0)} \quad but \quad g^*g_*(C_{x_0}) \neq C_{x_0}. \tag{5.12}$$

Proof. This results from (5.11). Compare with (0.6)–(0.7).

Let $\widetilde{C}_{g(x_0)} : t \to \widetilde{\phi}(t)$ be a curve on $\widetilde{M}$ such that $\widetilde{\phi}(0) = g(x_o)$ and let $\widetilde{Z}$ be a vector field on $\widetilde{M}$. Denote τ_t as parallel transport associated to the connection on M. We define parallel transport $\widetilde{\tau}_t$ on $\widetilde{M}$ as follows:

$$[\widetilde{\tau}_t^{-1}\widetilde{Z}(\widetilde{\phi}(t))]\widetilde{f} = E_\mu^g[[\tau_t^{-1}g^*(\widetilde{Z})(\psi(t))](\widetilde{f}og)], \tag{5.13}$$

where $\widetilde{C}_{g(x_0)} : t \to \widetilde{\phi}(t)$and $g_*\widetilde{C}_{g(x_0)} : t \to \psi(t)$.

We have $\widetilde{Z}(\widetilde{\phi}(t)) \in T_{\widetilde{\phi}(t)}$ and $\widetilde{\tau}_t^{-1} : T_{\widetilde{\phi}(t)} \to T_{g(x_0)}$, also $g(\psi(t)) = \widetilde{\phi}(t)$ and $\tau_t^{-1} : T_{\psi(t)} \to T_{x_0}$.

Lemma. *The connection on $\widetilde{M}$ associated to this parallel transport (5.13) has for covariant derivative (5.3).*

Proof. Let $\widetilde{\phi}(t)$ and $\psi(t)$ as above. By definition,

$$\widetilde{\nabla}_{\widetilde{X}}\widetilde{Y}(\xi) = \lim_{t\to 0} \frac{\widetilde{\tau}_t^{-1}\widetilde{Y}(\widetilde{\phi}(t)) - \widetilde{Y}(\widetilde{\phi}(0))}{t}$$

Let $Y = g^*(\widetilde{Y})$. Then $g'(x)Y(x) = \widetilde{Y}(g(x))$ and $Y(x)$ is in the orthogonal of $\ker g'(x)$. Since $\widetilde{\phi}(t) = g(\psi(t))$, and (5.13), we have

$$\widetilde{\tau}_t^{-1}\widetilde{Y}(\widetilde{\phi}(t))\widetilde{f} = [\widetilde{\tau}_t^{-1}g'(\psi(t))Y(\psi(t))]\widetilde{f} = E_\mu^g[(\tau_t^{-1}Y(\psi(t)))(x_0)(\widetilde{f}og)].$$

Thus, using (6.2) Part 1, we get

$$\lim_{t\to 0}\frac{\widetilde{\tau}_t^{-1}\widetilde{Y}(\widetilde{\phi}(t)) - \widetilde{Y}(\widetilde{\phi}(0))}{t} = \lim_{t\to 0}\frac{1}{t}(E_\mu^g[(\tau_t^{-1}Y(\psi(t))(\widetilde{f}og)] - Y(\psi(0))(\widetilde{f}og))$$

$$= E_\mu^g[\nabla_X Y(\widetilde{f}og)] = g_*(\nabla_{g^*(X)}g^*(Y))\widetilde{f}.$$

Remark. The map $\widetilde{\tau}^{-1}$ does not always preserve the metric in the tangent bundle $T(\widetilde{M})$. Let $\widetilde{\phi}(t) = g(\widetilde{\psi}(t))$; by (5.13),

$$J = (\widetilde{\tau}^{-1}[\widetilde{Z}(\widetilde{\phi}(t))]|\widetilde{\tau}^{-1}[\widetilde{Y}(\widetilde{\phi}(t))])_{T_{\widetilde{\phi}(0)}(\widetilde{M})}$$

$$= (g_*(\tau_t^{-1}g^*(\widetilde{Z})|g_*(\tau_t^{-1}g^*(\widetilde{Y}))_{T_{g(\psi(0))}(\widetilde{M})}.$$

Since the connection on M is Riemannian, we have

$$K = E_\mu^g[(\tau_t^{-1}g^*(\widetilde{Z})|\tau_t^{-1}g^*(\widetilde{Y}))_{T_{\psi(0)}(M)}]$$

$$= E_\mu^g[(g^*(\widetilde{Z})|g^*(\widetilde{Y}))_{T_{\psi(t)}(M)}] = (g_*g^*(\widetilde{Z})|g_*g^*(\widetilde{Y}))_{T_{\widetilde{\phi}(t)}(\widetilde{M})}$$

$$= (\widetilde{Z}|\widetilde{Y})_{T_{\widetilde{\phi}(t)}(\widetilde{M})},$$

where we use $g_* g^* = Id$ in the last equality. In general, $J \neq K$, thus the image connection of a Riemannian connection is not necessarily Riemannian. See 7 below for an example.

Proposition. *For a connection with covariant derivative $\nabla_Y X$, we denote ∇' the covariant derivative of the associated connection. That is,*

$$\nabla'_Y X = \nabla_X Y - [X, Y] \tag{5.14}$$

then for the image connections, we have

$$(g_* \nabla)' = g_*(\nabla'). \tag{5.15}$$

Proof. By definition of the image connection, and by (5.14), we have

$$(g_* \nabla')_{\widetilde{Y}} \widetilde{Z} = g_*(\nabla'_{g^*(\widetilde{Y})} g^*(\widetilde{Z})) = g_*(\nabla_{g^*(\widetilde{Z}} g^*(\widetilde{Y})) - g_*([g^*(\widetilde{Z}), g^*(\widetilde{Y})]). \tag{5.16}$$

We verify that the Lie bracket of vector fields satisfies

$$g_*([g^*(\widetilde{Z}), g^*(\widetilde{Y})] = [\widetilde{Z}, \widetilde{Y}].$$

Thus for the expression (5.16), we obtain

$$(g_* \nabla)_{\widetilde{Z}} \widetilde{Y} - [\widetilde{Z}, \widetilde{Y}] = (g_* \nabla)'_{\widetilde{Y}} \widetilde{Z}.$$

6 Example 1: Calculus of the density of the image measure

Let $g : R^2 \to R$ be given by $g(x, y) = x^2 - y^2$, and consider the gaussian probability measure $\mu = \frac{1}{2\pi} \exp(-\frac{(x^2+y^2)}{2}) dx dy$ on R^2. Then $g'(x, y) = (2x, -2y)$ is surjective except at $(x, y) = 0$. The following lemma is a consequence of (3.6).

Lemma. *Let $\widetilde{Z} = \alpha(\xi) \frac{d}{d\xi}$ be a vector field on R. Then*

$$Z = g^*(\widetilde{Z}) = \frac{\alpha(x^2 - y^2)}{2(x^2 + y^2)} (x \frac{\partial}{\partial x} - y \frac{\partial}{\partial y}) \tag{6.1}$$

Proof. Since $\ker g'(x_o, y_o)$ is the set of $(x, y) \in R^2$ such that $x_o x - y_o y = 0$, the orthogonal of $\ker g'(x_o, y_o)$ is the set of (x, y) such that $x y_o + y x_o = 0$, i.e., (x, y) is colinear to $(x_o, -y_o)$. Thus $Z = g^*(\widetilde{Z})$ is in the orthogonal of $\ker g'(x, y)$. Next, we verify that $g'(x, y) Z(x, y) = \widetilde{Z}(g(x, y))$. We put $\nu(x, y) = \frac{\alpha(x^2-y^2)}{2(x^2+y^2)}$; then

$$(g'(x, y) Z) \widetilde{\phi}(g(x, y)) = Z(\widetilde{\phi} o g)(x, y) = \nu(x, y)(x \frac{\partial}{\partial x} - y \frac{\partial}{\partial y})[\widetilde{\phi}(x^2 - y^2)]$$

$$= \nu(x,y)(2x^2+2y^2)\frac{d}{d\xi}\widetilde{\phi}(\xi) = \alpha(\xi)\frac{d}{d\xi}\widetilde{\phi}(\xi).$$

Proposition. *The density of the image measure with respect to the Lebesgue measure on M is a Bessel function. We have $g_*\mu = k(z)dz$ with*

$$k(z) = \frac{1}{2\pi}\int_0^{+\infty} e^{-(u^2+\frac{z^2}{u^2})}\frac{du}{u} = \frac{1}{2\pi}K_0(\frac{z}{2}), \tag{6.2}$$

where K_0 is a Bessel function (See Watson p. 183). Moreover, let $\phi : R^2 \to R$; then

$$(g_*\phi)(z) = \frac{1}{k(z)}\int_{-\infty}^{\infty}\phi(\frac{1}{2}(u+\frac{z}{u}), \frac{1}{2}(u-\frac{z}{u}))e^{-\frac{1}{4}(u^2+\frac{z^2}{u^2})}\frac{du}{4\pi|u|}. \tag{6.3}$$

7 Example 2: We show that the image of a Riemannian connection is not necessarily Riemannian for the image metric

We consider on R^3 a zero curvature connection defined by a frame. We put

$$b(x,y,\theta) = \begin{pmatrix} 1 & 0 & 0 \\ 0 & 1 & 0 \\ -y & x & 1 \end{pmatrix} \tag{7.1}$$

We consider the connection on R^3 whose covariant derivative is given by

$$\nabla_X Y(x,y,\theta) = \frac{d}{dt}_{|t=0} b(x,y,\theta)b(\phi(t))^{-1}Y(\phi(t)), \tag{7.2}$$

where $\phi(t)$ is a curve on R^3 such that $\phi(0) = (x,y,\theta)$ and $\frac{d}{dt}_{|t=0}\phi(t) = X(x,y,\theta)$. We verify that (7.2) defines a connection.

On R^3, we choose the metric a given by $g_{ij} =^{\dagger} (b^{-1})b^{-1}$ where ${}^{\dagger}(b^{-1})$ is the matrix transposed of b^{-1}. We denote by g^{ij} the matrix inverse of g_{ij}. Then

$$g_{ij} = \begin{pmatrix} 1+y^2 & -xy & y \\ -xy & 1+x^2 & -x \\ y & -x & 1 \end{pmatrix} \quad \text{and} \quad g^{ij} = \begin{pmatrix} 1 & 0 & -y \\ 0 & 1 & x \\ -y & x & 1+x^2+y^2 \end{pmatrix} \tag{7.3}$$

We have $\det g_{ij} = 1$ and the connection ∇ given by (7.2) is Riemannian for this metric. The Riemannian volume is $dv = dxdyd\theta$.

Let $Y = \alpha\frac{\partial}{\partial x} + \beta\frac{\partial}{\partial y} + \delta\frac{\partial}{\partial\theta}$. From (4.3) and since $\det g_{ij} = 1$, we deduce that

$$div_{dv}(Y) = \frac{\partial\alpha}{\partial x} + \frac{\partial\beta}{\partial y} + \frac{\partial\delta}{\partial\theta}. \tag{7.4}$$

For $\phi(t) = (x(t), y(t), \theta(t))$,

$$b(\phi(0))(b(\phi(t)))^{-1} = \begin{pmatrix} 1 & 0 & 0 \\ 0 & 1 & 0 \\ -(y(0) - y(t)) & x(0) - x(t) & 1 \end{pmatrix}.$$

Let $Z_1 = \frac{\partial}{\partial x}$, $Z_2 = \frac{\partial}{\partial y}$, $Z_3 = \frac{\partial}{\partial \theta}$, the coordinate vector fields. By taking $\phi_1(t) = (x+t, y, \theta)$, we deduce that for any vector field $Y = \alpha\frac{\partial}{\partial x} + \beta\frac{\partial}{\partial y} + \delta\frac{\partial}{\partial \theta}$, we have

$$\nabla_{Z_1}(Y) = \frac{d}{dt}_{|t=0} \begin{pmatrix} 1 & 0 & 0 \\ 0 & 1 & 0 \\ 0 & -t & 1 \end{pmatrix} Y(x+t, y, \theta)$$

Since $Y(x+t, y, \theta) = \alpha(x+t, y, \theta)Z_1 + \beta(x+t, y, \theta)Z_2 + \delta(x+t, y, \theta)Z_3$, we have

$$\nabla_{Z_1}(Y) = \frac{\partial \alpha}{\partial x}Z_1 + \frac{\partial \beta}{\partial x}Z_2 + \frac{\partial \delta}{\partial x}Z_3 - \beta Z_3. \tag{7.5}$$

Similarly, with $\phi_2(t) = (x, y+t, \theta)$,

$$\nabla_{Z_2}(Y) = \frac{d}{dt}_{|t=0} \begin{pmatrix} 1 & 0 & 0 \\ 0 & 1 & 0 \\ t & \grave{a} & 1 \end{pmatrix} Y(x, \dot{y}+t, \theta) = \frac{\partial \alpha}{\partial y}Z_1 + \frac{\partial \beta}{\partial y}Z_2 + \frac{\partial \delta}{\partial y}Z_3 + \alpha Z_3 \tag{7.6}$$

and $\phi_3(t) = (x, y, \theta + t)$ gives

$$\nabla_{Z_3}(Y) = \frac{d}{dt}_{|t=0} \begin{pmatrix} 1 & 0 & 0 \\ 0 & 1 & 0 \\ 0 & 0 & 1 \end{pmatrix} Y(x, y, \theta+t) = \frac{\partial \alpha}{\partial \theta}Z_1 + \frac{\partial \beta}{\partial \theta}Z_2 + \frac{\partial \delta}{\partial \theta}Z_3. \tag{7.7}$$

We deduce from (7.5)–(7.6)–(7.7)–(7.4) that

$$\sum_{i=1}^{3} \nabla_{Z_i}(Y)^i = div_{dv}(Y). \tag{7.8}$$

By (7.5)–(7.6)–(7.7), we have $\nabla_{Z_1}Z_1 = 0$, $\nabla_{Z_1}Z_2 = -Z_3$, $\nabla_{Z_1}Z_3 = 0$, $\nabla_{Z_2}Z_1 = Z_3$, $\nabla_{Z_2}Z_2 = 0$, $\nabla_{Z_2}Z_3 = 0$, $\nabla_{Z_3}Z_1 = \nabla_{Z_3}Z_2 = \nabla_{Z_3}Z_3 = 0$; thus $\gamma_{12}^3 = -1$ and $\gamma_{21}^3 = 1$. The other Christoffel symbols are zero. This implies $\sum_j \gamma_{ij}^j = 0$ and $\sum_j \gamma_{ji}^j = 0$ for any i. Thus, the torsion condition $\sum_i T_{ij}^i = 0$ (see (6.23) Part I) is satisfied. We also have $\mathcal{B}_\gamma(Y) = \mathcal{A}_\gamma(Y) = div_{dv}(Y)$.

Let $g : R^3 \to R^2$ be given by

$$g(x, y, \theta) = (x, y + \epsilon\theta). \tag{7.9}$$

On R^3, let the probability measure

$$\mu = \exp(-\frac{(x^2 + y^2 + \theta^2)}{2})\frac{dxdyd\theta}{(2\pi)^{3/2}} \tag{7.10}$$

then

$$g_*\mu = \exp(-\frac{x^2}{2})\exp(-\frac{u^2}{2(\epsilon^2+1)})\frac{dxdu}{2\pi\sqrt{\epsilon^2+1}} \tag{7.11}$$

If $\phi : R^3 \to R$, then

$$E^g_\mu(\phi(x,y,\theta)) = \frac{\sqrt{1+\epsilon^2}}{\epsilon\sqrt{2\pi}}\int_{-\infty}^{+\infty}\phi(x,y,\frac{u-y}{\epsilon})\exp(-\frac{1}{2}(\frac{1+\epsilon^2}{\epsilon^2})(y-\frac{u}{\epsilon^2+1})^2)dy \tag{7.12}$$

In particular, we have

$$E^g_\mu(y) = \frac{u}{1+\epsilon^2} \quad \text{and} \quad E^g_\mu(y^2) = 1 + \frac{u^2}{(1+\epsilon^2)^2} \tag{7.13}$$

For the image metric $b = g_*a$, we calculate b_{ij} with (4.5). Thus $(b^{-1})_{ij} = E^g_\mu[\lambda_{ij}]$ with

$$\lambda_{ij} = \begin{pmatrix} \frac{\partial g_i}{\partial c_1} & \frac{\partial g_i}{\partial c_2} & \frac{\partial g_i}{\partial c_3} \end{pmatrix} (a^{-1})_{kp} \begin{pmatrix} \frac{\partial g_j}{\partial c_1} \\ \frac{\partial g_j}{\partial c_2} \\ \frac{\partial g_j}{\partial c_3} \end{pmatrix} \tag{7.14}$$

where $g_1(x,y,\theta) = x$ and $g_2(x,y,\theta) = y + \epsilon\theta$.

$$(b^{-1})_{ij} = \begin{pmatrix} 1 & -\epsilon u(1+\epsilon^2)^{-1} \\ -\epsilon u(1+\epsilon^2)^{-1} & (1+\epsilon x)^2 + \epsilon^2 + u^2\epsilon^2(1+\epsilon^2)^{-2} \end{pmatrix}$$

$$= E^g_\mu \begin{pmatrix} 1 & -\epsilon y \\ -\epsilon y & (1+\epsilon x)^2 + \epsilon^2(1+y^2) \end{pmatrix} \tag{7.15}$$

For the image metric b_{ij} defined above, we verify that (4.1) and (4.4) are satisfied. On $\widetilde{M} = R^2$, let $\widetilde{Z} = \alpha\frac{\partial}{\partial x} + \beta\frac{\partial}{\partial u}$; then $g^*(\widetilde{Z}) = \alpha V_1 + \beta V_2$ with

$$V_1 = g^*(\frac{\partial}{\partial x}) = \frac{\partial}{\partial x} + \frac{\epsilon y(1+\epsilon x)}{\epsilon^2 + (1+\epsilon x)^2}(\frac{\partial}{\partial y} - \frac{1}{\epsilon}\frac{\partial}{\partial\theta}) \tag{7.16}$$

$$V_2 = g^*(\frac{\partial}{\partial u}) = \frac{1+\epsilon x}{\epsilon^2 + (1+\epsilon x)^2}(\frac{\partial}{\partial y} - \frac{1}{\epsilon}\frac{\partial}{\partial\theta}) + \frac{1}{\epsilon}\frac{\partial}{\partial\theta}, \tag{7.17}$$

where we calculate the components of V_2 with the identity

$$\frac{1+\epsilon x}{\epsilon^2 + (1+\epsilon x)^2} + \epsilon\frac{x + \epsilon(1+x^2)}{\epsilon^2 + (1+\epsilon x)^2} = 1.$$

Since $[\frac{\partial}{\partial x}, \frac{\partial}{\partial u}] = 0$ and $[Z_1, Z_2] \neq 0$, we have (see (0.8))

$$[g^*(\frac{\partial}{\partial x}), g^*(\frac{\partial}{\partial u})] \neq g^*([\frac{\partial}{\partial x}, \frac{\partial}{\partial u}]). \tag{7.18}$$

Let $Y_1 = \frac{\partial}{\partial x}$ and $Y_2 = \frac{\partial}{\partial u}$ be the coordinates vector fields on $\widetilde{M} = R^2$. For any vector field $\widetilde{Z}$ on $\widetilde{M} = R^2$, we verify that $g_*[g^*(\widetilde{Z})(\log|\det \lambda_{iq}|)] = \widetilde{Z}[\log|\det(b^{-1})_{ij}|]$ and we conclude by (3.6)–(7.16)–(7.17) that (see (3.13))

$$g_*[g^*(\widetilde{Z})(\log|\det(g^*(Y_i)|g^*(Y_j))_a|)] = \widetilde{Z}(\log|\det(Y_i|Y_j)_{g_*a}|). \tag{7.19}$$

Let $Y = \widetilde{A}\frac{\partial}{\partial x} + \widetilde{B}\frac{\partial}{\partial y} + \widetilde{C}\frac{\partial}{\partial \theta}$ and $Z = A\frac{\partial}{\partial x} + B\frac{\partial}{\partial y} + C\frac{\partial}{\partial \theta}$. We assume moreover that A and $\widetilde{A}$ are constants and that B, C, $\widetilde{B}$, $\widetilde{C}$ do not depend on θ. We deduce from (7.5)–(7.6)–(7.7) that for the covariant derivative of the connection (7.2),

$$\nabla_Y Z = (\widetilde{A}\frac{\partial B}{\partial x} + \widetilde{B}\frac{\partial B}{\partial y})\frac{\partial}{\partial y} + \widetilde{A}(\frac{\partial C}{\partial x} - B)\frac{\partial}{\partial \theta} + \widetilde{B}(\frac{\partial C}{\partial y} + A)\frac{\partial}{\partial \theta} \tag{7.20}$$

We consider the connection (7.2). We put $V_1 = g^*(\frac{\partial}{\partial x})$ and $V_2 = g^*(\frac{\partial}{\partial u})$. From (7.16)–(7.17)–(7.20), we see that

$$\nabla_{V_1} V_1 = \epsilon^3 y(\epsilon^2 + (1+\epsilon x)^2)^{-2}(\epsilon\frac{\partial}{\partial y} - \frac{\partial}{\partial \theta})$$

$$\nabla_{V_1} V_2 = (\frac{d}{dx}\frac{1+\epsilon x}{\epsilon^2 + (1+\epsilon x)^2})(\frac{\partial}{\partial y} - \frac{1}{\epsilon}\frac{\partial}{\partial \theta}) - \frac{1+\epsilon x}{\epsilon^2 + (1+\epsilon x)^2}\frac{\partial}{\partial \theta}$$

$$\nabla_{V_2} V_1 = \frac{(1+\epsilon x)}{\epsilon^2 + (1+\epsilon x)^2}\frac{\partial}{\partial \theta} + (1+\epsilon x)^2(\epsilon^2 + (1+\epsilon x)^2)^{-2}(\epsilon\frac{\partial}{\partial y} - \frac{\partial}{\partial \theta})$$

$$\nabla_{V_2} V_2 = 0 \tag{7.21}$$

Let $H(x,u) = \alpha(x,u)\frac{\partial}{\partial x} + \beta(x,u)\frac{\partial}{\partial u}$ and $W(x,u) = \widetilde{\alpha}(x,u)\frac{\partial}{\partial x} + \widetilde{\beta}(x,u)\frac{\partial}{\partial u}$; then

$$\nabla_{g^*(W)} g^*(H) = K(\frac{\partial}{\partial y} - \frac{1}{\epsilon}\frac{\partial}{\partial \theta}) + (\alpha\widetilde{\beta} - \widetilde{\alpha}\beta)\frac{(1+\epsilon x)}{\epsilon^2 + (1+\epsilon x)^2}\frac{\partial}{\partial \theta} \tag{7.22}$$

with

$$K = (\alpha\widetilde{\beta} - \widetilde{\alpha}\beta)\epsilon(1+\epsilon x)^2(\epsilon^2 + (1+\epsilon x)^2)^{-2} \quad + \epsilon^3\widetilde{\alpha}(\alpha\epsilon y + \beta)(\epsilon^2 + (1+\epsilon x)^2)^{-2}. \tag{7.23}$$

On $\widetilde{M} = R^2$, let $\widetilde{\gamma}$ be the image connection with covariant derivative $\widetilde{\nabla}$ defined by (5.3). With (7.21) and (15.3), we prove

Lemma. *On R^2, we put $Y_1(x,u) = \frac{\partial}{\partial x}$ and $Y_2(x,u) = \frac{\partial}{\partial u}$. For the image connection $\widetilde{\gamma}$ on $\widetilde{M} = R^2$, we have*

$$\widetilde{\nabla}_{Y_1} Y_1 = 0 \quad , \widetilde{\nabla}_{Y_1} Y_2 = -\frac{\epsilon(1+\epsilon x)}{\epsilon^2 + (1+\epsilon x)^2} Y_2$$

$$\widetilde{\nabla}_{Y_2} Y_1 = \frac{\epsilon(1+\epsilon x)}{\epsilon^2 + (1+\epsilon x)^2} Y_2 \quad , \widetilde{\nabla}_{Y_2} Y_2 = 0 \tag{7.24}$$

For the image connection $\widetilde{\gamma}$, the torsion condition $\mathcal{T} = 0$ (see (6.23) Part I) is not satisfied since

$$\mathcal{T}(Y_1) = -\frac{2\epsilon(1+\epsilon x)}{\epsilon^2 + (1+\epsilon x)^2} Y_2 \tag{7.25}$$

However, the condition $\mathcal{T} = 0$ was valid for the initial connection on R^3. See (7.8).

With (7.16), we give an example for (0.20): We take on R^2 the vector field $Y_1(x,u) = \frac{\partial}{\partial x}$, we see that $g^*(\widetilde{\nabla}_{\widetilde{Y}_1}\widetilde{Y}_1) \neq \nabla_{V_1} V_1$.

Proposition. *The image connection $\widetilde{\gamma}$ given by (7.24) is not Riemannian for the image metric (7.15). However it satisfies the condition $\mathcal{B}_{\widetilde{\gamma}}(Y) = div_{d\widetilde{v}}(Y)$ where $d\widetilde{v}$ is the volume element on R^2 with the image metric (7.15). See(6.23) Part 1.*

Proof. For the image connection, we call $\widetilde{\gamma}_{ij}^k$ the Christoffel symbols in the basis Y_1, Y_2, we have

$$\widetilde{\nabla}_{Y_i} Y_j = \widetilde{\gamma}_{ij}^k Y_k$$

where

$$\widetilde{\gamma}_{11} = \widetilde{\gamma}_{22} = 0 \qquad , \quad \widetilde{\gamma}_{12} = \begin{pmatrix} 0 \\ -\epsilon(1+\epsilon x)(\epsilon^2 + (1+\epsilon x)^2)^{-1} \end{pmatrix} = -\widetilde{\gamma}_{21}$$

If the connection is Riemannian, we have $\partial_k b^{ij} = -b^{lj}\widetilde{\gamma}_{kl}^i - b^{li}\widetilde{\gamma}_{kl}^j$, where (b^{ij}) is the matrix given by (7.15). We see that

$$\frac{\partial}{\partial x} b^{12} = 0 \quad \text{but} \quad -b^{21}\widetilde{\gamma}_{12}^2 \neq 0;$$

thus the image connection is not Riemannian for the image metric though the initial connection on R^3 was Riemannian.

With (6.23) Part I, we verify that the image connection $\widetilde{\gamma}$ satisfies the condition $div_{dv}(\widetilde{Y}) = \mathcal{B}_{\widetilde{\gamma}}(\widetilde{Y})$, though it is not Riemannian. Consider the associated connection $\widetilde{\gamma}' = g_*(\gamma)'$, see (5.15) Part II. For γ' the connection associated to γ, we have $div_{dv}(Y) = \mathcal{B}_{\gamma'}(Y)$ but for the image connection $\widetilde{\gamma}'$, $div_{d\widetilde{v}}(\widetilde{Y}) \neq \mathcal{B}_{(g_*\gamma')}(\widetilde{Y})$.

III Functors on the tangent bundle

1 Extended vector fields

Let $T_x(M)$ be the tangent space at $x \in M$ and $u \in T_x(M)$. For a vector field X on M, we consider the differential equation

$$\frac{d}{dt} U_t(x) = [U_t(x), X(x)] \tag{1.1}$$

with the initial condition $U_0(x) = u$. In the chart c, we have $X = \sum_i (\alpha_i o c) Z_i^c$ and $U_t(x) = \sum_i u_i(t,x) Z_i^c$.

From (1.11) Part I,

$$[U_t(x), X(x)] = \sum_{i,j}[u_i(t,x)\frac{\partial}{\partial c_i}\alpha_j - \alpha_i\frac{\partial}{\partial c_i}u_j(t,x)]Z_j^c. \tag{1.2}$$

Thus (1.1) is equivalent to a system of partial differential equations. We assume that $U_t(x)$ is a solution of this system.

Definition. Let X be a vector field on M; let $\phi(t)$ be a curve on M such that $\phi(0) = x$ and $\frac{d}{dt}\phi(t) = X(\phi(t))$. We define a vector field X^{ext} on the tangent bundle $T(M)$: for a function $f : T(M) \to R$ and (x, u) with $u \in T_x(M)$, we put

$$X^{ext_1} f(x,u) = \frac{d}{dt}_{|t=0} f(\phi(t), U_t(\phi(t))) \tag{1.2a}$$

where $U_t(x)$ is a vector field solution of (1.1).

Lemma. *In a chart c, let* $X = \sum_i (\alpha_i o c) Z_i^c$, *and consider the function* f *from* $T(M)$ *to* R *be given in local coordinates by*

$$f : (c_1, c_2, \ldots, c_n, u_1, u_2, \ldots, u_n) \to f(c_1, c_2, \ldots, c_n, u_1, u_2, \ldots, u_n)$$

We have

$$(X^{ext_1} f)(x,u) = \sum_i \alpha_i \frac{\partial f}{\partial c_i} + \sum_j (\sum_i u_i \frac{\partial \alpha_j}{\partial c_i}) \frac{\partial f}{\partial u_j} \tag{1.3}$$

Proof.

$$\frac{d}{dt}_{|t=0} f(\phi(t)), U_t(\phi(t))) = \sum_i (\frac{\partial f}{\partial c_i})\alpha_i + \sum_j \frac{\partial f}{\partial u_j}[u_i \frac{\partial \alpha_j}{\partial c_i} - \frac{\partial u_j}{\partial c_i}\alpha_i] + \frac{\partial f}{\partial u_j}\frac{\partial u_j}{\partial c_i}\alpha_i, \tag{1.4}$$

where

$$\frac{d}{dt}_{|t=0} U_t(\phi(t)) = \frac{d}{dt}_{|t=0} U_t(x) + \frac{d}{dt}_{|t=0} U_0(\phi(t)). \tag{1.5}$$

Definition. Let X be a vector field on M, and $(x, u) \in T(M)$; we put

$$X^{ext_2} f(x,u) = \frac{d}{dt}_{|t=0} f(x, u + tX(x)). \tag{1.6}$$

In local coordinates, assume that $X = \sum_i (\alpha_i o c) Z_i^c$, we have for (1.6)

$$X^{ext_2} f(x,u) = \sum_i \alpha_i \frac{\partial f}{\partial u_i}. \tag{1.7}$$

Notation. Let $(x, u) \in T(M)$, in a neighborhood of x. We can define a vector field U such that $U(x) = u$. In local coordinates, since $u \in T_x(M)$, we have $u = \sum_i u_i Z_i^c(x)$; we put

$$U = \sum_i u_i Z_i^c. \tag{1.8}$$

Lemma. *Let $h : M \to R$ and X be a vector field on M; then*

$$(hX)^{ext_1}(x,u) = h(x)X^{ext_1}(x,u) + Uh(x)X^{ext_2}(x,u)$$

$$(hX)^{ext_2}(x,u) = h(x)X^{ext_2}(x,u), \tag{1.9}$$

where U is given by (1.8).

Proof. We express (1.9) in local coordinates using (1.3) and (1.7).

Lemma. *Let X and Y be two vector fields on M. For the Lie bracket, we have*

$$[X^{ext_1}, Y^{ext_1}] = [X,Y]^{ext_1}$$

$$[X^{ext_1}, Y^{ext_2}] = [X,Y]^{ext_2} \quad , \qquad [X^{ext_2}, Y^{ext_2}] = 0.$$

Proof. We verify the formulas using the local coordinates. The extended vector fields defined by (1.2) and (1.6) are independent of any connection on M. Given a connection γ on M, we can define an extension of vector fields which is related to γ.

Definition. Let γ be a connection on M and let X be a vector field on M. We give a curve $\phi(t)$ on M such that $\phi(0) = x$ and $\frac{d}{dt}\phi(t) = X(\phi(t))$. Let $\tau_t : T_{\phi(0)}(M) \to T_{\phi(t)}(M)$ be the associated parallel transport along $\phi(t)$. For $u \in T_x(M)$, we consider $\tau_t u$. Let $f : T(M) \to R$. We put

$$X^{ext(\gamma)} f(x,u) = \frac{d}{dt}_{|t=0} f(\phi(t).\tau_t u) \tag{1.10}$$

In local coordinates, let γ^i_{jk} be the Christoffel symbols associated to the connection γ; the vectors $u(t) = \tau_t u$ are solutions of the differential system

$$\frac{d}{dt}u_i(t) = -\gamma^i_{jk}(\phi(t))\frac{d}{dt}\phi_j(t)u_k(t). \tag{1.11}$$

For $X = \sum_i (\alpha_i o c) Z^c_i$, we have

$$(X^{ext(\gamma)}) f(x,u) = \sum_i \alpha_i \frac{\partial f}{\partial c_i} - \sum_{ijk} \gamma^i_{jk}(\phi(t))\alpha_j u_k \frac{\partial f}{\partial u_i}. \tag{1.12}$$

Let c be a chart in a neighborhood of $x \in M$, and let $(x,u) \in T(M)$; we define a basis of $T_{(x,u)}(T(M))$ as follows: We take the coordinates vector fields extended by (1.3), i.e., $(Z^c_i)^{ext_1}$, $i = 1, \dots, n$ and the coordinates vector fields extended by (1.6), i.e., $(Z^c_i)^{ext_2}$, $i = 1, \dots, n$. Then

$$(Z^c_i)^{ext_1}(x,u) \quad , \qquad (Z^c_i)^{ext_2}(x,u) \qquad i = 1, \dots, n \tag{1.13}$$

form a basis of $T_{(x,u)}(T(M))$.

2 Riemannian metric on the tangent bundle. Extension of a measure on M to a measure on $T(M)$. Image of extended vector fields

Given an extension of vector fields, either (1.3), (1.6) or (1.12), and a Riemannian metric a on M, we define a Riemannian metric a_T on $T(M)$ by following [6]. In local coordinates, we put

$$\big((Z_i^c)^{ext}(x,u)|((Z_j^c)^{ext}(x,u)\big)_{a_T} = a_{ij}(x) \quad , \quad \big((Z_i^c)^{ext}(x,u)|V_i(x,u)\big)_{a_T} = 0$$

$$\big(V_i(x,u)|V_j(x,u)\big)_{a_T} = a_{ij}(x). \tag{2.1}$$

For this metric, we have

$$\det(a_T)_{ij}(x,u) = (\det a_{ij}(x))^2. \tag{2.2}$$

This metric defines a Riemannian volume dv_T on $T(T(M))$. Let c be a chart of M, we consider the chart (c, dc) on $T(M)$; then

$$(c, dc)_* dv_T = (\det a_{ij}(c_1, \dots, c_n)) dc_1 dc_2 \dots dc_n du_1 du_2 \dots du_n \tag{2.3}$$

where $dc_1 dc_2 \dots dc_n du_1 du_2 \dots du_n$ is the Lebesgue measure on R^{2n}. Let μ be a probability measure on M. we may extend μ into a probability measure μ^{ext} on $T(M)$; for that, we consider $T(M)$ as $M \times R^n$ and on R^n, we can take the Gaussian measure induced by the metric on the tangent space $T_m(M) = R^n$.

Let M and $\widetilde{M}$, be two differentiable manifolds and $g : M \to \widetilde{M}$ be a differentiable map as in Part II, it induces a differentiable map $(g, dg) : T(M) \to T(\widetilde{M})$

$$(g, dg)(x, u) = (g(x), g'(x)u). \tag{2.4}$$

Given a function f on $T(M)$, we define the direct image of f as the conditional expectation

$$(g, dg)_* f = E_{\mu^{ext}}^{(g,dg)}(f). \tag{2.5}$$

With the notations of Part II, we conjecture that

$$(g_* X)^{ext} = (g, dg)_*(X^{ext}). \tag{2.6}$$

References

[1] H. Airault, Projection of the infinitesimal generator of a diffusion, *Journal of Funct. Anal.* **85** (1989), 353–391.

[2] B.K. Driver, A Cameron-Martin quasi-invariance theorem for brownian motion on a compact manifold, *Journal of Functional Anal.* **110** (1992), 603–608.

[3] N. Ikeda and S. Watanabe, *Stochastic differential equations and diffusion processes*, North-Holland/Kodansha, Volume 24, 1981.

[4] S. Kobayashi and K. Nomizu, *Foundations of Differential Geometry*, Vol.1, John Wiley and Sons, New York, 1963.

[5] P. Malliavin, *Stochastic Analysis*, Springer, 1997.

[6] S. Sasaki, On the differential geometry of tangent bundles of Riemannian manifolds, Selected papers Edited by Shun-ichi Tachibana, Kinokuniya Company Ltd., Tokyo, 1985.

H. Airault
INSSET, Université de Picardie Jules Verne
48 rue Raspail
02100 Saint-Quentin (Aisne) France
e-mail:`hairaultinsset.u-picardie.fr`

P. Malliavin
10 rue Saint Louis en l'Isle
75004, Paris, France
e-mail: `sliccr.jussieu.fr`

Stochastic Volterra Equations with Singular Kernels

L. Coutin and L. Decreusefond

1 Introduction

Motivated by potential applications to fractional Brownian motion ([3]), we study Volterra stochastic differential of the form

$$X_t = x + \int_0^t K(t,s)b(s,X_s)ds + \int_0^t K(t,s)\sigma(s,X_s)\,dB_s, \qquad \text{(E)}$$

where $(B_s,\ s \in [0,1])$ is a one-dimensional standard Brownian motion and $(K(t,s),\ t,s \in [0,1])$ is a deterministic kernel whose properties will be made precise below but for which we do not assume any boundedness property.

Actually, when σ is a constant and K is given by (2.2), we obtain

$$X_t = x + \int_0^t K(t,s)b(s,X_s)ds + \sigma W_t^H,$$

where W^H is the fractional Brownian motion of Hurst parameter H — see the example below. In this particular case the main feature is that K is highly singular as a kernel but the integral map canonically associated to it, i.e.,

$$Kf(t) = \int_0^t K(t,s)f(s)\,ds,$$

is a regularizing operator. This explains why we work as much as possible with the properties of the map K and not with those of the kernel $K(t,s)$. The problem is then in the treatment of the stochastic integral. One of the main difficulties is to control (Hölder) regularity with respect to t of the stochastic integral on the right-hand side of (E). This has been the object of a previous paper [2], the hypothesis of which we simplify here. With the result obtained in that paper, the proof of existence and uniqueness of the solution of (E) is achieved as usual by a fixed point technique.

However, some other problems arise when we study the Gross–Sobolev regularity of the solution of (E) and the expression of its derivative. They are two kinds of problems. On the one hand, the singularity of the kernel and on the other hand, the fact that the Gross–Sobolev derivative of the solution is the solution of a linear but time-dependent stochastic differential equation. We eventually give a somewhat explicit expression for the Gross–Sobolev derivative of the solution — in this article, the approach owes much to the ideas developed in [5].

Note that the specific form of the drift ensures both a symmetric role to b and σ and the existence of weak solutions to (E); see [1] for the application of this notion to the nonlinear filtering theory with fractional Gaussian noise. For other stochastic differential equations related to fractional Brownian motion, we refer to [7, 8, 16].

The equations we have to deal with are of Volterra type, but our work does not seem to be subsumed by previous articles on this subject (see for instance [6, 11, 12]) because our kernel is weakly regular and we are looking for classical solutions and not distribution-valued ones. Our work is only done in one dimension but it is straightforward to extend it to higher dimensions.

This article is organized as follows: in the next section, we show how hypothesis (A) is sufficient to entail those of the main theorem of [2]. As an example, we address the case of fractional Brownian motion. In Section 3, we prove the main theorem of existence and uniqueness of the solution of (E). In Section 4, we prove that under an extra boundedness assumption on σ this solution is Gross–Sobolev differentiable, and we give an integral representation of it.

2 Preliminaries

Consider a measurable kernel $(K(t,s), s, t \in [0,1])$ and denote also by K the (formal) linear map

$$Kf(t) = \int_0^t K(t,s) f(s)\, ds.$$

HYPOTHESIS (A). *We assume throughout that there exists $\gamma > 0$ such that K is continuous both from $\mathcal{L}^1([0,t])$ into $\mathcal{I}_{\gamma,2}([0,t])$ and from $\mathcal{L}^2([0,t])$ into $\mathcal{I}_{\gamma+1/2,2}([0,t])$ for any $t \in [0,1]$.*

Theorem 2.1. *Let u be an adapted process belonging to $L^r(\Omega \times [0,1], P \otimes dt)$. If $r \geq 2$, the family of random variables $(M_t(u) = \int_0^t K(t,s) u_s\, dB_s,\ t \in [0,1])$ has a version whose trajectories belong almost surely to $\mathcal{I}_{\gamma-\epsilon,r}$ for any $\epsilon \in (0,\gamma)$. Consequently, if $\gamma > 1/r$, the sample paths are almost surely $(\gamma - 1/r - \epsilon)$-Hölder continuous for any $\epsilon \in (0,\gamma)$. Moreover, the following maximal inequality holds:*

$$\| M(u) \|_{L^r(\Omega; \mathcal{I}_{\gamma-\epsilon,r}([0,t]))} \leq c \| u \|_{L^r(\Omega \times [0,t])}. \tag{2.1}$$

Proof. It is sufficient to prove that hypotheses I to III of [2] are satisfied. Let $\delta = \gamma - \epsilon,$. We have to show that

1. K is continuous from $\mathcal{L}^2$ into $\mathcal{B}$, the space of bounded functions on $[0,1]$.
2. K and $K_\delta = I_{0+}^{-\delta} \circ K$ are Hilbert-Schmidt from $\mathcal{L}^2$ into itself,

Since $\gamma > 0$, it is well known (see section A of the appendix) that $\mathcal{I}_{\gamma+1/2,1/2} \subset \mathcal{H}_\gamma \subset \mathcal{B}$ and the first point follows.

Since $\gamma > 0$, the embedding of $\mathcal{I}_{\gamma+1/2,2}$ in $\mathcal{L}^2$ is Hilbert-Schmidt (see [14]) and so is K from $\mathcal{L}^2$ into itself. Moreover, K_δ is continuous from $\mathcal{L}^2$ into $\mathcal{I}_{1/2+\epsilon,2}$ and the embedding of $\mathcal{I}_{1/2+\epsilon,2}$ in $\mathcal{L}^2$ is Hilbert-Schmidt, hence K_δ is Hilbert–Schmidt from $\mathcal{L}^2$ into itself. □

As an example (in fact the motivating one), consider $K = K_H$, the kernel that is related to fractional Brownian motion. For any H in $(0, 1)$, the fractional Brownian motion of index (Hurst parameter) H, $\{W_t^H;\ t \in [0, 1]\}$ is the unique centered Gaussian process whose covariance kernel is given by

$$R_H(s,t) = \mathrm{E}[W_s^H W_t^H] = \frac{V_H}{2}\left(s^{2H} + t^{2H} - |t-s|^{2H}\right)$$

where

$$V_H = \frac{\Gamma(2-2H)\cos(\pi H)}{\pi H(1-2H)}.$$

It has been proved (see [3]) that there exists a standard Brownian motion such that almost surely

$$W_t^H = \int_0^t K_H(t,s)\, dB_s,$$

where

$$K_H(t,r) = \frac{(t-r)^{H-\frac{1}{2}}}{\Gamma(H+\frac{1}{2})} F(\frac{1}{2} - H, H - \frac{1}{2}, H + \frac{1}{2}, 1 - \frac{t}{r}) 1_{[0,t)}(r). \tag{2.2}$$

The Gauss hypergeometric function $F(\alpha, \beta, \gamma, z)$ (see [10]) is the analytic continuation on $\mathbb{C} \times \mathbb{C} \times \mathbb{C}\backslash\{-1, -2, \dots\} \times \{z \in \mathbb{C}, Arg|1-z| < \pi\}$ of the power series

$$\sum_{k=0}^{+\infty} \frac{(\alpha)_k(\beta)_k}{(\gamma)_k k!} z^k.$$

Here $(\alpha)_k$ denotes the Pochhammer symbol defined by

$$(a)_0 = 1 \text{ and } (a)_k = \frac{\Gamma(a+k)}{\Gamma(a)} = a(a+1)\dots(a+k-1).$$

It is well-known from [13] that K_H maps continuously $\mathcal{L}^p$ into $\mathcal{I}_{H+1/2,p}$ for any $p \geq 1$, so in this case, hypothesis (A) is fulfilled for any $\gamma < H$ since $\mathcal{I}_{H+1/2,1} \subset \mathcal{I}_{H-\epsilon,2}$ for any ϵ sufficiently small.

3 Existence and uniqueness of the solution of (E)

In the following, we denote by c any irrelevant constant appearing in the computations.

Definition 3.1. By a solution of the equation (E), we mean a real-valued, progressively measurable stochastic process $X = \{X_t,\ t \in I\}$ such that X belongs to $L^2(\Omega \times [0, 1], \mathrm{P} \otimes dt)$, and for any t, X_t is a.s. a solution of (E).

In the next section r will denote a fixed real strictly greater than $\max(2, \gamma^{-1})$.

Theorem 3.2. *Let b and σ be L-Lipschitz continuous with respect to their second variable, and uniform with respect to their first variable: For all t in* $[0, 1]$, *for all x, y in* $\mathbf{R}$,

$$|b(t, x) - b(t, y)| + |\sigma(t, x) - \sigma(t, y)| \le L|x - y|.$$

Assume also that there exist x_0 and y_0 in $\mathbb{R}$, such that $b(., x_0)$ and $\sigma(., y_0)$ belong to $\mathcal{L}^r$. The differential equation (E) has then a unique continuous solution that belongs to $\mathcal{H}_{\gamma-1/r-\epsilon}$, for any $\epsilon \in (0, \gamma)$.

Proof. The proof proceeds as usual by a fixed point technique. It is sufficient to note that for any u and v two progressively measurable processes belonging to $L^r(\Omega \times [0, 1], \mathrm{P} \otimes dt)$, we have the following:

1. The process $\int_0^t K(t, s)u_s ds + \int_0^t K(t, s)v_s\, dB_s$ is continuous (according to (A) and Theorem 2.1) and adapted, hence progressively measurable;
2. Since $\mathcal{I}_{\gamma+1/2,2}$ is continuously embedded in $\mathcal{B}$, according to hypothesis (A), for any $1/r < \delta < \gamma$,

$$\mathrm{E}[\sup_{s \le t} |\int_0^t K(t, s)u_s\, ds|^r] \le \mathrm{E}[\|\int_0^{.} K(., s)u_s\, ds\|_{\mathcal{I}_{\delta,r}}^r] \tag{3.1}$$

$$\le c\mathrm{E}[\int_0^t |u_s|^r\, ds]; \tag{3.2}$$

3. According to Theorem 2.1, for any $1/r < \delta < \gamma$,

$$\mathrm{E}[\sup_{s \le t} |\int_0^t K(t, s)v_s\, dB_s|^r] \le c\mathrm{E}[\|\int_0^{.} K(., s)v_s\, dB_s\|_{\mathcal{I}_{\delta,r}}^r] \tag{3.3}$$

$$\le c\, \|v\|_{L^r(\Omega\times[0,1])}^r. \tag{3.4}$$

Uniqueness is then a consequence of (3.2), (3.4) and the Gronwall lemma. According to (3.2), (3.4), the Picard sequence defined by

$$X_t^0 = x,\ X_t^n = x + \int_0^t K(t, s)b(s, X_s^{n-1})\, ds + \int_0^t K(t, s)\sigma(s, X_s^{n-1})\, dB_s$$

is a Cauchy sequence in $L^r(\Omega \times [0,1], \mathrm{P} \otimes dt)$. We denote by X its limit. It is clearly a continuous and adapted solution of (E). Furthermore, inequalities (3.1) and (3.3) entail that the convergence also holds in $L^r(\Omega; \mathcal{I}_{\delta,r})$, so the solution has a.s. $(\delta - 1/r)$-Hölder continuous sample paths. □

Remark. Note that in the case of the fBm of Hurst index $H < 1/2$, we cannot work as usual in L^2 but only in $L^{1/H}$, hence we have to ensure stronger regularity on the coefficients, i.e., $b(., x_0)$ and $\sigma(., x_0)$ must belong to $\mathcal{L}^{1/H}$.

By the same techniques, we can prove

Theorem 3.3. *Under the hypothesis of the previous theorem, the map which sends x to the solution of (E) with initial condition x is continuous from $\mathbb{R}$ in $\mathcal{L}^r(\Omega \times [0,1], P \otimes dt)$.*

4 Gross–Sobolev regularity of X

We are now interested in the Gross–Sobolev differentiability of X_t.

Lemma 4.1. *For u adapted, belonging to $\mathcal{L}^2(\Omega \times [0,1]; \mathbb{D}_{2,1})$, for any $t \in [0,1]$, the distribution $\nabla(\int_0^t K(t,s) u_s \, dB_s)$ exists as an $L^2(\Omega \times [0,t])$-random variable and satisfies*

$$\|\nabla(\int_0^t K(t,s) u_s \, ds)\|_{L^2(\Omega \times [0,t])} \le c \, \|u\|_{L^2(\Omega \times [0,t])} \tag{4.1}$$

Lemma 4.2. *For u bounded, adapted belonging to $\mathcal{L}^2(\Omega \times [0,1]; \mathbb{D}_{2,1})$, for any $t \in [0,1]$, the distribution $\nabla(\int_0^t K(t,s) u_s \, dB_s)$ exists as a $L^2(\Omega \times [0,t])$ random variable and satisfies*

$$\|\nabla(\int_0^t K(t,s) u_s \, dB_s)\|_{L^2(\Omega \times [0,t])} \le c \, \|u\|_{L^2(\Omega \times [0,t]; \mathbb{D}_{2,1})}. \tag{4.2}$$

Proof. We only prove (4.2) since inequality (4.1) is simpler to show and its proof proceeds along the same lines. According to (A), for any t, the random variable $\int_0^t K(t,s) u_s \, dB_s$ belongs to $L^r(\Omega, \mathrm{P})$ and thus has a derivative in the distributional sense. For any $\xi \in \mathbb{D}_\infty(\mathcal{L}^2)$,

$$\begin{aligned}
\langle \nabla(\int_0^t K(t,s) u_s \, dB_s), \, \xi \rangle_{\mathbb{D}_{-\infty}, \mathbb{D}_\infty} &= \mathrm{E}[\int_0^t K(t,s) u_s \, dB_s \, \delta(\xi)] \\
&= \mathrm{E}[\int_0^t K(t,s) u_s \, \xi_s \, ds] + \mathrm{E}[\int_0^t K(t,s) \Big(\int_0^t \nabla_r u_s \, \xi_r \, dr \Big) dB_s].
\end{aligned}$$

Hypothesis (A) induces that for any $\xi \in \mathbb{D}_\infty(\mathcal{L}^2)$, we have

$$\begin{aligned}
\mathrm{E}[| \int_0^t K(t,s) u_s \, \xi_s \, ds |] &\le c \, \|K(t,.) u\|_{L^2(\Omega \times [0,1])} \|\xi\|_{L^2(\Omega \times [0,1])} \\
&\le c \, \|u\|_\infty \|K^*(\epsilon_t)\|_{\mathcal{L}^2} \|\xi\|_{L^2(\Omega \times [0,1])}.
\end{aligned}$$

Since ϵ_t belongs to the dual of $\mathcal{I}_{\gamma+1/2,2}$, according to (A), we have

$$\|K^*(\epsilon_t)\|_{\mathcal{L}^2} \le c\,\|\epsilon_t\|_{\mathcal{I}^*_{\gamma+1/2,2}} = ct^{\gamma}. \tag{4.3}$$

According to Theorem 2.1 for $r=2$, since $\mathcal{I}_{\gamma-\epsilon,2} \subset \mathcal{L}^2$, for any $\xi \in \mathbb{D}_\infty(\mathcal{L}^2)$,

$$\begin{aligned} \mathrm{E}[|\int_0^t \xi_\tau \int_0^t K(t,s)\nabla_\tau u_s\,dB_s\,d\tau|] \\ \le c\,\|\xi\|_{L^2(\Omega\times[0,1])}\mathrm{E}[\int_0^t\int_0^s |\nabla_\tau u_s|^2\,d\tau\,ds]^{1/2}. \end{aligned} \tag{4.4}$$

The result is then a consequence of (4.3) and (4.4) and Proposition 3 of [15, page 37]. □

Remark. If $\gamma > 1/2$, inequalities (4.1) and (4.2) are truly uniform with respect to t, i.e., for instance,

$$\|\sup_{t\le T} \nabla(\int_0^t K(t,s)u_s\,dB_s)\|_{L^2(\Omega\times[0,T])} \le c\,\|u\|_{L^2(\Omega\times[0,T];\mathbb{D}_{2,1})}.$$

Theorem 4.3. *The hypotheses of Theorem 3.2 are assumed to hold. Moreover, b and σ are supposed to be once continuously differentiable with respect to their space variable, with bounded derivative; assume furthermore that σ is bounded. For any $t \in I$, the value at t of the solution of Equation* (E)*, denoted by X_t, belongs to $\mathbb{D}_{2,1}$. For any $\xi \in \mathbb{H}$,*

$$\begin{aligned} < X_t, \xi >_{\mathbb{H}} = \int_0^t K(t,s)\sigma(s,X_s)\xi_s\,ds \\ + \int_0^t K(t,u)\frac{\partial b}{\partial x}(u,X_u) < \nabla X_u, \xi >_{\mathbb{H}}\,du \\ + \int_0^t K(t,u)\frac{\partial \sigma}{\partial x}(u,X_u) < \nabla X_u, \xi >_{\mathbb{H}}\,dB_u. \end{aligned} \tag{4.5}$$

Moreover, for $\xi \in L^r(\Omega\times[0,1])$, $(< \nabla X_t, \xi >_{\mathbb{H}},\ t\in[0,1])$ belongs to $L^r(\Omega\times[0,1])$.

Proof. Let X^n be the Picard sequence already defined in the proof of Theorem 3.2. Using Lemmas 4.1 and 4.2, we prove by induction on n that X^n belongs to $\mathbb{D}_{2,1}$ and that

$$\|X_t^{n+1} - X_t^n,\|^2_{\mathbb{D}_{2,1}} \le c\int_0^t \|X_t^n - X_t^{n+1}\|^2_{\mathbb{D}_{2,1}}\,ds \le \frac{(ct)^n}{n!}x^2.$$

It follows that $\sup_n \|X_t^n\|_{\mathbb{D}_{2,1}}$ is finite and thus that there exists a weakly convergent subsequence in $\mathbb{D}_{2,1}$. Since X_t^n converges to X_t in $L^2(\Omega)$, the closability of

∇ entails that X_t belongs to $\mathbb{D}_{2,1}$. Since for any n,

$$
\begin{aligned}
< X_t^n, \xi >_{\mathbb{H}} = & \int_0^t K(t,s)\sigma(s, X_s^{n-1})\xi_s \, ds \\
& + \int_0^t K(t,u)\frac{\partial b}{\partial x}(u, X_u^{n-1}) < \nabla X_u^{n-1}, \xi >_{\mathbb{H}} du \\
& + \int_0^t K(t,u)\frac{\partial \sigma}{\partial x}(u, X_u^{n-1}) < \nabla X_u^{n-1}, \xi >_{\mathbb{H}} dB_u,
\end{aligned}
$$

a straightforward application of the dominated convergence theorem yields to (4.5). Since $\mathcal{L}^r$ is continuously imbedded in $\mathcal{L}^2$ and $\mathcal{I}_{\gamma+1/2,2}$ is continuously embedded in $\mathcal{C}_0([0,1];\mathbb{R})$, we have

$$
\begin{aligned}
\mathrm{E}[\sup_{s\le t} | < \nabla X_s, \xi >_{\mathbb{H}} |^r] & \le c\Big(\mathrm{E}[\int_0^1 |\xi_s|^r \, ds] + \mathrm{E}[\int_0^t | < \nabla X_s, \xi >_{\mathbb{H}} |^r \, ds]\Big) \\
& \le c\Big(\mathrm{E}[\int_0^1 |\xi_s|^r \, ds] + \mathrm{E}[\int_0^t \sup_{u\le s} | < \nabla X_u, \xi >_{\mathbb{H}} |^r \, ds]\Big).
\end{aligned}
$$

By Gronwall lemma, it follows that $< \nabla X_., \xi >_{\mathbb{H}}$ belongs to $L^r(\Omega \times [0,1])$. □

Theorem 4.4. *Assume that the hypothesis of Theorem 4.3 holds. For any* $\xi \in L^r(\Omega \times [0,1])$, *the equation*

$$
\begin{aligned}
Y_t = < K(t,.)\sigma \circ X, \xi >_{\mathbb{H}} & + \int_0^t K(t,u)\frac{\partial b}{\partial x}(u, X_u)Y_u \, du \\
& + \int_0^t K(t,u)\frac{\partial \sigma}{\partial x}(u, X_u)Y_u \, dB_u \quad (4.6)
\end{aligned}
$$

has one and only one solution belonging to $L^r(\Omega \times [0,1])$.

Proof. Theorem 4.3 states that the process $(\langle \nabla X_t, \xi \rangle_{\mathbb{H}}, t \in [0,1])$ is a solution of (4.6) with the desired integrability property.

If Y and Z are two such solutions, according to hypothesis (A) and to Theorem 2.1, we have

$$
\mathrm{E}[|Y_t - Z_t|^r] \le c \int_0^t \mathrm{E}[|Y_s - Z_s|^r] \, ds.
$$

By iteration, this induces that $Y = Z$, $\mathrm{P} \otimes dt$ almost everywhere. □

Theorem 4.5. *Assume that the hypothesis of Theorem 4.3 holds. Let* $V_0(t,s)$ *be a measurable deterministic kernel such that*

$$
\int_0^1 \int_0^1 |V_0(u,s)|^r \, du \, ds < \infty. \quad (4.7)
$$

For $n \geq 1$, consider

$$V_{n+1}(t,s) = \int_s^t K(t,u)\frac{\partial b}{\partial x}(u, X_u)V_n(u,s)\,du + \int_s^t K(t,u)\frac{\partial \sigma}{\partial x}(u, X_u)V_n(u,s)\,dB_u.$$

The two following properties hold:

V1 *$L(t,s) = \sum_{n=0}^{+\infty} V_n(t,s)$ is a convergent series in $L^r(\Omega \times [0,1]^2)$.*

V2 *$L(t,s)$ is a solution of*

$$L(t,s) - V_0(t,s) = \int_s^t K(t,u)\frac{\partial b}{\partial x}(u, X_u)L(u,s)\,du + \int_s^t K(t,u)\frac{\partial \sigma}{\partial x}(u, X_u)L(u,s)\,dB_u. \quad (4.8)$$

Proof. By the techniques used above, we can show that

$$\int_0^\zeta \int_0^1 |V_{n+1}(t,s)|^r\,dt\,ds \leq c \int_0^\zeta \int_0^1 \int_s^t |V_n(u,s)|^r\,du\,ds\,dt,$$

hence if we set

$$\psi_n(t) = \int_0^t \int_0^1 |V_n(u,s)|^r\,ds\,du,$$

the previous equation reads as

$$\psi_{n+1}(\zeta) \leq c \int_0^\zeta \psi_n(t)\,dt.$$

Applying the Gronwall lemma and (4.7), it follows that the series $\sum_n V_n$ is convergent in $L^r(\Omega \times [0,1]^2)$. Moreover, it is straightforward that

$$\int_s^t K(t,u)\frac{\partial b}{\partial x}(u, X_u)\sum_{j=0}^{n} V_j(u,s)\,du + \int_s^t K(t,u)\frac{\partial \sigma}{\partial x}(u, X_u)\sum_{j=0}^{n} V_j(u,s)\,dB_u = \sum_{j=0}^{n+1} V_j(t,s) - V_0(t,s).$$

According to hypothesis (A) and Theorem 2.1, the left-hand side of the last equation converges to the right-hand side of (4.8). □

Hypothesis (B). *We assume that there exists an almost surely positive function g such that $V_0(t,s) = K(t,s)g(s)$ satisfies (4.7).*

Theorem 4.6. [Parameter variation formula] *Assume that all hypotheses made so far hold. Consider the space $\mathbb{H}_0$ of elements of $\mathbb{H}$ satisfying $\xi g^{-1} \in \mathcal{L}^r$. Let L be defined as in the previous theorem with the value of V_0 taken in hypothesis (B). For any $\xi \in \mathbb{H}_0$, let $Y_t = \int_0^t L(t,s)\sigma(X_s)g^{-1}(s)\xi(s)\,ds$. For any $t \in [0,1]$, we have $< \nabla X_t, \xi >_{\mathbb{H}} = Y_t$, P almost surely.*

Proof. For $\xi \in \mathbb{H}_0$, since σ is bounded, Y belongs to $L^r(\Omega \times [0,1])$. It is clear that Y is a formal solution of (4.6) hence by theorem 4.4, the equality $< \nabla X_t, \xi >_{\mathbb{H}} =$ Y_t follows. □

Remark. For fractional Brownian motion, it is proved in [3] that

$$0 \le K_H(t,s) \le c\,(t-s)^{H-1/2} s^{-|H-1/2|},$$

hence hypothesis (B) is satisfied with $g(s) = s^{|H-1/2|}$. It is a little counter-intuitive that as H increases towards 1, hypothesis (B) requires an increasing value of ν to be fulfilled. It is due to the increasing singularity of $K_H(t,s)$, whereas K_H as a map is more and more regularizing.

A. Deterministic fractional calculus

For $f \in \mathcal{L}^1([0,1])$, the left and right fractional integrals of f are defined by

$$(I^\alpha_{0+} f)(x) = \frac{1}{\Gamma(\alpha)} \int_0^x f(t)(x-t)^{\alpha-1} dt\ ,\ x \ge 0,$$

$$(I^\alpha_{b-} f)(x) = \frac{1}{\Gamma(\alpha)} \int_x^b f(t)(t-x)^{\alpha-1} dt\ ,\ x \le b,$$

where $\alpha > 0$ and $I^0 = Id$. In what follows, T is a real in $(0,1]$. For any $\alpha \ge 0$, any $f \in \mathcal{L}^p([0,T])$ and $g \in \mathcal{L}^q([0,T])$ where $p^{-1} + q^{-1} \le \alpha$, we have

$$\int_0^T f(s)(I^\alpha_{0+} g)(s)\,ds = \int_0^T (I^\alpha_{T-} f)(s) g(s)\,ds. \qquad (A.15)$$

The Besov space $I^\alpha_{0+}(\mathcal{L}^p([0,T])) = \mathcal{I}_{\alpha,p}([0,T])$ is usually equipped with the norm

$$\|f\|_{\mathcal{I}_{\alpha,p}} = \|I^{-\alpha}_{0+} f\|_{\mathcal{L}^p([0,T])}.$$

We then have the following continuity results (see [4, 13]):

Proposition A.1 *For each* $0 < T \le 1$,

1. *If $0 < \alpha < 1$, $1 < p < 1/\alpha$, then I^α_{0+} is a bounded operator from $\mathcal{L}^p([0,T])$ into $\mathcal{L}^q([0,T])$ with $q = p(1-\alpha p)^{-1}$.*

2. *For any* $0 < \alpha < 1$ *and any* $p \geq 1$, $\mathcal{I}_{\alpha,p}([0,T])$ *is continuously embedded in* $\mathcal{H}_{\alpha-1/p}([0,T])$ *provided that* $\alpha - 1/p > 0$. $\mathcal{H}_{\nu}([0,T])$ *denotes the space of Hölder-continuous functions, null at time* 0, *equipped with the usual norm*

$$\| f \|_{\mathcal{H}_{\nu}([0,T])} = \sup_{0 \leq t \neq s \leq T} \frac{|f(t) - f(s)|}{|t-s|^{\nu}}.$$

By $I_{0+}^{-\alpha}$, respectively $I_{1-}^{-\alpha}$, we mean the inverse map of I_{0+}^{α}, respectively I_{1-}^{α}. When we do not make precise the interval $[0,T]$ in the notation of $\mathcal{L}^p$ spaces or of Besov spaces, it is intended that $T = 1$.

B. Malliavin calculus

We only give the few results needed; further details can be found in [9, 15]. We work on the standard Wiener space $(\Omega, \mathbb{H}, \mathrm{P})$ where Ω is the Banach space of continuous functions from $[0,1]$ into $\mathbb{R}$, null at time 0, equipped with the sup-norm. $\mathbb{H}$ is the Hilbert space of absolutely continuous function with the norm $\|h\|_{\mathbb{H}} = \|\dot{h}\|_{\mathcal{L}^2}$, where $\dot{h}$ is the time derivative of h. A mapping ϕ from Ω into some separable Hilbert space X is called cylindrical if it is of the form $\phi(w) = f(< v_1, w >, \cdots, < v_n, w >)$ where $f \in C_0^{\infty}(\mathbb{R}^n, X)$ and $v_i \in \Omega^*$ for $i = 1, \ldots, n$. For such a function we define $\nabla\phi$ as

$$\nabla\phi(w) = \sum_{i=1}^{n} \partial_i f(< v_1, w >, \cdots, < v_n, w >)\tilde{v}_i \,,$$

where $\tilde{v}_i$ is the image of v_i under the injection $\Omega^* \hookrightarrow \mathcal{L}^2$. From the quasi-invariance of the Wiener measure, it follows that ∇ is a closable operator on $L^p(\Omega; X)$, $p \geq 1$, and we will denote its closure with the same notation. The powers of ∇ are defined by iterating this procedure. For $p > 1$, $k \in \mathbb{N}$, we denote by $\mathbb{D}_{p,k}(X)$ the completion of X-valued cylindrical functions under the norm

$$\|\phi\|_{p,k} = \sum_{i=0}^{k} \|\nabla^i \phi\|_{L^p(\Omega; X \otimes (\mathcal{L}^2)^{\otimes i})} \,.$$

Let us denote by δ the formal adjoint of ∇ with respect to Wiener measure. A classical result states that δ is an extension of the Itô integral, and thus we have

$$\mathrm{E}[\int_0^t u_s \, dB_s \, \varphi] = \mathrm{E}[\int_0^t u_s \nabla_s \varphi \, ds] \qquad (B.16)$$

for any u adapted in $L^2(\Omega; \mathcal{L}^2)$ and any $\varphi \in \mathbb{D}_{2,1}$, where $\{B_t = \delta(\mathbf{1}_{[0,t]}), t \in [0,1]\}$ is a standard Brownian motion on (Ω, P).

References

[1] L. Coutin and L. Decreusefond, Abstract Non-linear Filtering Theory in the Presence of Fractional Brownian Motion. *Annals Applied Probability*, **9**:4 (1999), 1058–1090.

[2] L. Decreusefond, Regularity properties of some stochastic Volterra integrals with singular kernel, to appear in *Potential Analysis*.

[3] L. Decreusefond and A.S. Üstünel, Stochastic Analysis of Fractional Brownian Motion. *Potential Analysis*, **10**:2 (1999), 177–214.

[4] D. Feyel and A. de La Pradelle, On Fractional Brownian Processes. *Potential Analysis*, **10**:3 (1999), 273–288.

[5] F. Hirsch, Propriété d'absolue continuité pour les équations différentielles stochastiques dépendant du passé, *Journal of Functional Analysis*, **76**:1 (1988), 193–216.

[6] M. Lewin, Weak Solutions to Volterra's Population Equation with Diffusion and Noise, In: *Integral methods in science and engineering, Proceedings of the international conference, ISME '93*, C. Constanda, ed., 1994, pp. 163–172.

[7] S.J. Lin, Stochastic analysis of fractional Brownian motions, *Stochastics and Stochastics Reports*, **55**:1-2 (1995), 121–140.

[8] T. Lyons, Differential equations driven by rough signals, I. An extension of an inequality of L. C. Young, *Mathematical Research Letters*, **4** (1994), 451–464.

[9] Paul Malliavin, *Stochastic Analysis*, Grundlehren der Mathematischen Wissenschaften, 313. Springer, Heidelberg, 1997.

[10] A.F. Nikiforov and V.B. Uvarov, *Special Functions of Mathematical Physics*, Birkhäuser, 1988.

[11] E. Pardoux and P. Protter, Stochastic Volterra Equations with Anticipating Coefficients, *Annals of Probability*, **18**:4 (1990), 1635–1656.

[12] P. Protter, Volterra Equations driven by Semimartingales, *Annals of Probability*, **13**:2 (1985), 519–531.

[13] S.G. Samko, A.A. Kilbas, and O.I. Marichev, *Fractional Integrals and Derivatives*, Gordon & Breach, 1993.

[14] L. Schwartz, editor, *Applications radonifiantes*, École Polytechnique, 1970.

[15] A.S. Üstünel, *An Introduction to Analysis on Wiener Space*, Vol. 1610, Lectures Notes in Mathematics, Springer, Heidelberg, 1995.

[16] M. Zähle, Integration with respect to Fractal Functions and Stochastic Calculus, *Probability Theory and Related Fields*, 111, 1998.

L. Coutin
Département de Statistiques et Probabilités
Université Paul Sabatier
118, route de Narbonne
31062 Toulouse cedex, France
email: coutin@cict.fr
and
L. Decreusefond
E.N.S.T.
46, rue Barrault
75634 Paris, cedex 13, France
E-mail: decreuse@enst.fr

Stochastic Diffeology and Homotopy

Rémi Léandre

1 Introduction

There are two stochastic cohomology theories of the loop space: In the first, we endow the loop space with the Brownian bridge measure if the underlying manifold is Riemannian. Sobolev spaces of forms are defined because a Hilbert tangent space is used, although the Brownian loop is only continuous ([8], [2]). To a random form is associated a series of numbers, that is its Sobolev norms, and the stochastic exterior derivative acts continuously over the intersection of Sobolev spaces ([10], [11], [13]). It is shown that if the manifold is simply connected, then the Sobolev stochastic cohomology of forms in this sense (a metric invariant) is equal to the Hochschild cohomology of forms over the manifold ([11], [13]) and therefore to the de Rham cohomology of the smooth loop space (a differentiable invariant). This result is a stochastic generalization of a result of Adams ([1]) and Chen ([4]), which says that the Hochschild cohomology is equal to the cohomology of smooth loops.

There are no Sobolev structures in the second theory involved with stochastic cohomology groups. This theory is based on a stochastic generalization of the notion of plots of Chen ([4]) and Souriau ([20]), which leads to a calculus in the set of semimartingales. There are at least two stochastic diffeologies associated to the loop space. For the first, we consider the uniform topology over the loop space, and a very large class of semimartingales is allowed to describe the stochastics plots ([12]). In the second ([14]), we consider Hölderian topology, such that there is a partition of unity associated to a special cover of the loop space, and we consider a smaller class of semimartingales that is allowed in order to describe the stochastic plots. Since there are partitions of unity associated to the Hölderian topology, the stochastic cohomology associated to this second diffeology (or to a system of stochastic plots) is equal to the de Rham cohomology of the Hölderian loop space. Therefore the stochastic cohomology associated to this restricted class of stochastic plots is invariant by differentiable homotopies of the manifold.

The goal of this article is the following: if we consider the stochastic cohomology associated to the large system of plots of [12], we get cohomology groups that are invariant under differentiable homotopies of the manifold. The proof is direct, and does not suppose an explicit computation of the stochastic cohomology groups in terms of cohomology groups of some deterministic objects, which are invariant by homotopy. (Let us recall that in [12], there is a definition of stochastic homology groups of the loop space that is computed in terms of the deterministic homology groups of the differentiable loop space). This comes from the fact if $X_.$ is a semimartingale over the manifold M and if F is a smooth function from M into M, $F(X_.)$ is still a semimartingale over M by the Itô formula. For the Brownian bridge, F should be an isometry.

This work gives a stochastic example and a stochastic interpretation to the considerations of the annexes A, B, C of the work of P. Iglesias ([7]).

2 Lift of a finite dimensional application

Let (Ω, F_s, P) $s \in [0, 1]$ be a filtered probability space. Let M be a smooth manifold of dimension d_M embedded into $R^{d_{1,M}}$.

Let us recall the definition of a stochastic plot ([12]):

Definition 2.1. A stochastic plot from an open subset U of R^{d_2} into $L_x(M)$ is given by the following data:

(a) A countable family of applications ϕ_i from U into the set of continuous semimartingales over $R^{d_{1,M}}$, starting from x in M, such that the quadratic variation $< \phi_i(u), \phi_i(v) >$ depends smoothly on all L^p on u and v in U; moreover, we suppose that the finite variational part of $\phi_i(u)$ depends smoothly on all L^p for the finite variational norm.

(b) A countable measurable partition Ω_i of Ω such that, almost surely over Ω_i, for all u in U, $\phi_i(u) \in L_x(M)$.

We denote by $(U, \phi_i, \Omega_i, L_x(M))$ this stochastic plot. We identify two stochastic plots $(U, \phi_i^1, \Omega_i^1, L_x(M))$ and $(U, \phi_j^2, \Omega_j^2, L_x(M))$ if $\phi_i^1 = \phi_j^2$ almost surely over $\Omega_i^1 \cap \Omega_j^2$.

Let $f : M \to N$ be a smooth function from M into N such that $f(x) = y$. We extend it to a smooth application from $R^{d_{1,M}}$ into $R^{d_{1,N}}$ with bounded derivatives of all orders.

Proposition 2.2. *Let $(U, \phi_i, \Omega_i, L_x(M))$ a stochastic plot from U into $L_x(M)$. Then $(U, f \circ \phi_i, \Omega_i, L_y(N))$ is a stochastic plot from U into $L_y(N)$.*

Proof. This comes from the Itô formula:

$$\begin{aligned} f(\phi_i(u)_t) = f(x) + \int_0^t f'(\phi_i(u)_s)\delta\phi_i(u)_s \\ + \int_0^t \sum \frac{\partial^2 f}{\partial y_j \partial y_k}(\phi_i(u)_s) d < \phi_i(u), \phi_i(u) > j, k, s \end{aligned} \tag{2.1}$$

δ denotes the Itô integral. Since the f have bounded derivatives over $R^{d_{1,M}}$, assumption (a) is checked. Clearly, (b) is checked. ◇

Remark. By the equivalence relation that identifies two plots, it follows that the stochastic plot $(U, f \circ \phi_i, \Omega_i, L_y(N))$ does not depend on the way we extend f from M to $R^{d_{1,M}}$.

Let us recall the definition of a stochastic form over $L_x(M)$.

Definition 2.3. An n-form σ_{st} over the loop space $L_x(M)$ is given by the following:

(a) To each stochastic plot $(U, \phi_i, \Omega_i, L_x(M))$, we associate almost surely over each Ω_i a random smooth n form $\sigma_U(\phi_i)$ over U.

(b) Let j be a smooth application from U^1 into U^2 and let there be the stochastic plot $(U^1, \phi_i^2, \Omega_i, L_x(M))$. We have almost surely the identity of random forms:

$$\sigma_{U^1}(\phi_i^2 \circ j) = j^* \sigma_{U^2}(\phi_i^2)$$

(c) Let $(U, \phi_i^1, \Omega_i^1, L_x(M))$ and $(U, \phi_j^2, \Omega_j^2, L_x(M))$ be two stochastics plots such that there exists i and j and a measurable transformation $\psi : \Omega_i^1 \to \Omega_j^2$ over a set of strictly positive probability such that

$$\phi_j^2 \circ \psi = \phi_i^1 \tag{2.2}$$

Then

$$\sigma_U(\phi_j^2) \circ \psi = \sigma_U(\phi_i^1) \tag{2.3}$$

Remark. In (b) we perform a change of variable over the parameter set U. In (c) we perform a change of parameter over the probability space Ω_i.

Let $f : M \to N$ which transforms the based point x into the based point y. Let σ_{st} be a stochastic form over $L_y(N)$. We can define the pullback $f^*\sigma_{st}$ of this stochastic form.

Definition 2.4. Let σ_{st} be a stochastic form over $L_y(N)$. $f^*\sigma_{st}$ is defined by the stochastic formulas: Let $(U, \phi_i, \Omega_i, L_x(M))$ be a stochastic plot. $f^*\sigma_U(\phi_i) = \sigma_U(f \circ \phi_i)$.

It is clear that $f^*\sigma_U(\phi_i)$ satisfies all criteria of Definition 2.3.

The following proposition is clear:

Proposition 2.5. *Let $f_{2,1} : M_1 \to M_2$ and $f_{3,2} : M_2 \to M_3$ be two smooth applications such that $f_{2,1}(x_1) = x_2$ and such that $f_{3,2}(x_2) = x_3$. Then $(f_{3,2} \circ f_{2,1})^*$ applies a stochastic form over $L_{x_3}(M_3)$ to a stochastic form over $L_{x_1}(M_1)$. And we have*

$$(f_{3,2} \circ f_{2,1})^* = (f_{2,1})^* \circ (f_{3,2})^* \tag{2.4}$$

We would like this application f^* to work in the stochastic cohomology group. First, we have to define a stochastic exterior derivative.

Definition 2.6. Let σ_{st} be a stochastic form over $L_x(M)$. $d\sigma_{st}$ is the stochastic form defined by the systems of $d\sigma_U(\phi_i)$ for the stochastic plots $(U, \phi_i, \Omega_i, L_x(M))$. It is clear that the system of $d\sigma_U(\phi_i)$ checks the consistency relations (a) and (b) of Definition 2.3.

Proposition 2.7. *Let f be a smooth application from M into N such that $f(x) = y$. Let σ_{st}^N be a stochastic form over $L_y(M)$. We have:*

$$f^* d\sigma_{st}^N = df^* \sigma_{st}^N \tag{2.5}$$

Proof. $f^* d\sigma_{st}^N$ is defined by $(d\sigma_u^N)(f \circ \phi_i)$ if $(U, \phi_i, \Omega_i, L_x(M))$ is a stochastic plot over $L_x(M)$. $f^* \sigma_{st}^N$ is defined by $\sigma_U^N(f \circ \phi_i)$. Therefore 2.5. ◇

Definition 2.8. $H_{st}^*(L_x(M)) = Kerd/Imd$ are called the stochastic cohomology groups of $L_x(M)$.

We deduce from Proposition 2.7 and Proposition 2.5 the main theorem of this section.

Theorem 2.9. *f induces a linear map from $H_{st}^*(L_y(N))$ into $H_{st}^*(L_x(M))$ called f^*. Moreover,*

$$(f \circ g)^* = g^* \circ f^* \tag{2.6}$$

3 Invariance under homotopy of the cohomology groups

Let (M, x) be a based compact manifold. Let (N, y) be another based manifold.

Definition 3.1. An application $f : (M, x) \to (M, x)$ is homotopic to the identity if there exist a smooth application $F(., t)$ from

$$\begin{aligned} F(., 1) = & \quad f(.) \\ F(., 0) = & \quad I_d \\ F(x, t) = & \quad x \end{aligned} \tag{3.7}$$

Definition 3.2. (M, x) is said to be homotopically equivalent to (N, y) if there exists a smooth application $f : (M, x) \to (N, y)$ (i.e., $f(x) = y$) and a smooth application $g : (N, y) \to (M, x)$ (i.e., $g(y) = x$) such that $g \circ f : (M, x) \to (M, x)$ is homotopic to the identity and $f \circ g : (N, y) \to (N, y)$ is homotopic to the identity.

The main theorem of this paper is the following:

Theorem 3.3. *Let us suppose that (M, x) is homotopically equivalent to (N, y). Then $H_{st}^*(L_x(M)) = H_{st}^*(L_y(N))$.*

Proof. it is enough to show that $f^* \circ g^* = 1$ and that $g^* \circ f^* = 1$. For this, it suffices to show that $f : (M, x) \to (M, x)$ is homotopically equivalent to the identity, $f^* = 1$.

Let σ_{st} be a stochastic closed form over $L_x(M)$, and let $(U, \phi_i, \Omega_i, L_x(M))$ be a stochastic plot. We consider the extended plot $(U \times [0, 1], F(\phi_i, t), \Omega_i, L_x(M))$ where we have extended the function $F(., t)$ over $R^{d_{1,M}} \times [0, 1]$. We deduce that $\sigma_{U\times[0,1]}(F(\phi_i, .))$, a closed form. $\sigma_{U\times\{1\}}(F(\phi_i, 1)$ represents $f^*\sigma_{st}$. $\sigma_{U\times\{0\}}(F(\phi_i, 0))$ represents σ_{st}. Moreover, we have a retraction J_ϵ from $U\times[0, 1]$ into $U \times \{0\}$: $J_\epsilon(u, t) = (u, \epsilon t)$. Moreover,

$$\sigma_{U\times[0,1]} = J_0^*\sigma_{U\times[0,1]} + \int dJ_\epsilon^*\sigma_{U\times[0,1]}(\frac{\partial}{\partial\epsilon}J_\epsilon, .)d\epsilon \tag{3.8}$$

since $d\sigma_{U\times[0,1]}(F(\phi_i, .)) = 0$. We deduce that

$$\sigma_{U\times\{1\}}(f(\phi_i)) = \sigma_{U\times\{0\}}(\phi_i) + d_u \int_0^1 J_\epsilon^*\sigma_{U\times[0,1]}(\frac{\partial}{\partial\epsilon}J_\epsilon, .)d\epsilon \tag{3.9}$$

Therefore

$$\sigma_U(f(\phi_i)) = \sigma_U(\phi_i) + d_u \int_0^1 J_\epsilon^*\sigma_{U\times[0,1]}(F(\phi(., t))(\frac{\partial}{\partial\epsilon}J_\epsilon, .)d\epsilon \tag{3.10}$$

We must now check that the form $\tilde{\sigma}_U = \int_0^1 J_\epsilon^*(F(\phi_i, .))(\frac{\partial J_\epsilon}{\partial\epsilon}, .)d\epsilon$ represents a stochastic form. So we see that $\sigma_{U\times[0,1]}(F(\phi_i, t))(\frac{\partial J_\epsilon}{\partial\epsilon}, .)$ check properties (b) and (c) of Definition 2.3. Only property (b) is not clear: Let $j : U_1 \to U_2$ be a smooth application from U_1 into U_2. $F(\phi_i(j(u_2)), t)$ is a stochastic plot over $U_1 \times [0, 1]$. $J = (j, I_d)$ is the application from $U_1 \times [0, 1] \to U_2 \times [0, 1]$ which (u_1, t) associates to $(j(u_1), t)$. $\sigma_{U_2\times[0,1]}(F(\phi_i(u_2), t)) = \tilde{\sigma}_{U_2} + \overline{\sigma}_{U_2} \wedge dt$. We do over it the change of variable associates to J. We have

$$J^*(\tilde{\sigma}_{U_2} + \overline{\sigma}_{U_2} \wedge dt) = j^*\tilde{\sigma}_{U_2} + (j^*\overline{\sigma}_{U_2}) \wedge dt \tag{3.11}$$

This proves our assumption. Therefore, if σ_{st} is closed, $f^*\sigma_{st} = \sigma_{st} + d\tilde{\sigma}_{st}$. f^* is equal to the identity in the cohomology. ◇

References

[1] Adams J.F., On the cobar construction, *Proc. Nat. Acad. Sci. U.S.A.* **42**(1956), 346–373.

[2] Bismut J.M., *Large deviations and the Malliavin Calculus*, Progress in Math 45, Birkhäuser, 1984.

[3] Bott R., Tu L.W., *Differential forms in algebraic topology*, Springer, 1986.

[4] Chen K.T., Iterated path integrals of differential forms and loop space homology, *Ann. Math.*, **107** (1973), 213–237.

[5] Getzler E., Jones J.D.S., Petrack S., Differential forms on loop spaces and the cyclic bar complex, *Topology*, **30** (1991), 339–373.

[6] Iglesias P., Thesis, Université de Provence, 1985.

[7] Iglesias P., La trilogie du moment, *Ann. Institut Fourier*, **45**:3 (1995), 825–857.

[8] Jones J.D.S., Léandre R., L^p Chen forms on loop spaces, in *Stochastic Analysis*, M. Barlow, N. Bingham eds., Cambridge University Press, 1991, 104–162.

[9] Kusuoka S., De Rham Cohomology of Wiener-Riemannian manifolds. preprint, 1990.

[10] Léandre R., Cohomologie de Bismut–Nualart–Pardoux et cohomologie de Hochschild entière, in: *Séminaire de Probabilités XXX in honour of P.A. Meyer and J. Neveu*, J. Azéma, M. Emery, M. Yor, eds., Lectures Notes Math. 1626, 1996, 68–100.

[11] Léandre R., Brownian cohomology of an homogeneous manifold, in *New trends in Stochastic Analysis*, K.D. Elworthy, S. Kusuoka, I. Shigekawa, eds., World Scientific, 1997, 305–347.

[12] Léandre R., Singular integral homology of the stochastic loop space, *Infinite dimensional analysis, quantum probability and related topics*, **1**:1 (1998), 17–31.

[13] Léandre R., Stochastic Adams theorem for a general compact manifold, to appear *Reviews in Math. Physics*.

[14] Léandre R., Stochastic cohomology of Chen–Souriau and line bundle over the Brownian bridge, to appear *Probability Theory and Related Fields*.

[15] Léandre R., Stochastic plots and universal cover of the loop space, to appear *Potential Analysis*.

[16] Léandre R., Smolyanov O., Stochastic homology of the loop space, in: *Analysis on Infinite-Dimentional Lie Groups and Algebras*, H. Heyer, J. Marion, eds., World Scientific, 1999, 229–235.

[17] Ramer R., On the de Rham complex of finite codimensional forms on infinite dimensional manifold, Thesis, University of Warwick, 1974.

[18] Shigekawa I., De Rham–Hodge–Kodaira decomposition on an abstract Wiener space, *J. Math.*, Kyoto Univ. **26** (1986), 191–202.

[19] Smolyanov O., De Rham's current's and Stoke's formula in a Hilbert space, *Sov. Math. Dokl.*, **33** (1986), 140–144.

[20] Souriau J.M., Un algorithme générateur de structures quantiques, in: *Elie Cartan et les Mathématiques d'aujourd'hui*, Astérisque, 1985, 341–399.

Rémi Léandre
Institut Elie Cartan
Département de Mathématiques
Université Henri Poincaré
54000, Vandoeuvre-les-Nancy, France
email: leandre@iecn.u-nancy.fr

Some Results on Entropic Projections

C. Léonard

1 Introduction

We give a short survey of results related to the maximum entropy method. In this article we use the large deviations approach rather than the more direct convex analytical one. Indeed, the proposed applications are naturally stated in terms of large random particle systems, namely, the existence and construction problems for Schrödinger's bridges and Nelson's diffusion processes. These problems arise from probabilistic approaches to quantum mechanics.

Let R be a fixed reference probability measure. The optimization problem to be investigated is the minimization of the relative entropy $I(\cdot \mid R)$, subject to a general linear constraint ($-I(\cdot \mid R)$ is concave; it is *the* entropy to be maximized.

We introduce a constraint function φ which allows the description of a general infinite dimensional linear constraint. Our assumptions on φ are exponential integrability conditions with respect to R, called *Cramér's conditions*: the very strong one (2.5), the strong one (4.1), and the weak one (2.2).

In Section 2, the dual equality of our optimization problem is obtained under the very strong Cramér condition. It yields a criterion of existence of a unique minimizer: the *I-projection* (see Definition 2.1). In Section 3, this criterion is specified for the problems of existence of Schrödinger's bridges and Nelson's diffusion processes. In Section 4, we give a characterization of the I-projection, under the strong Cramér condition. This result definitely improves previous related results in the literature. In Section 5, we give a brief glimpse of the situation when φ is not very integrable: under the weak Cramér condition. In this situation, it may happen that the I-projection no longer exists. The minimization problem has to be replaced by an "extended" one. Its minimizers may no longer be unique and may admit a singular part that is not a measure (lack of σ-additivity). To keep our large deviations approach, we state an extension of Sanov's theorem in Theorem 5.1. The situation is illustrated with a simple example in Subsections 5.1 and 5.3. We call *entropic projection* the minimizers of the extended relative entropy — the rate function of the extended Sanov theorem (see Definition 5.3). It appears that I-projections are entropic projections.

Recall that a sequence $\{X_n\}_{n\geq 1}$ obeys the *large deviation principle* (LDP) in a topological space with the rate function I if for any measurable subset A :

$$
\begin{aligned}
-\inf_{x\in\mathrm{int}(A)} I(x) &\leq \liminf_{n\to\infty} \frac{1}{n}\log P(X_n \in A) \\
&\leq \limsup_{n\to\infty} \frac{1}{n}\log P(X_n \in A) \leq -\inf_{x\in\mathrm{cl}(A)} I(x)
\end{aligned}
$$

The rate function I is said to be a good rate function if it is inf-compact. For more details about large deviations, see the book of A. Dembo and O. Zeitouni [13].

2 Under the very strong Cramér condition

2.1 Sanov's theorem

Let $(Z_i)_{i\geq 1}$ be an independent identically distributed sequence of random elements with values in a measurable space $(\Omega, \mathcal{A})$ and common law $R \in \mathcal{P}(\Omega)$: the set of probability measures on $(\Omega, \mathcal{A})$. Sanov's theorem ([13], Theorem 6.2.10) states that the sequence of empirical measures $L_n = \frac{1}{n}\sum_{i=1}^n \delta_{Z_i} \in \mathcal{P}(\Omega)$ (δ stands for the Dirac measure) satisfies a LDP in $\mathcal{P}(\Omega)$ with the weak topology $\sigma(\mathcal{P}(\Omega), B)$, where B is the space of measurable bounded functions on Ω. Its good rate function is the relative entropy with respect to R, and it is given for any $P \in \mathcal{P}(\Omega)$ by

$$
\begin{aligned}
I(P \mid R) &= \sup_{f\in B}\left\{\int_\Omega f\,dR - \log\int_\Omega e^f\,dR\right\} \\
&= \begin{cases} \int_\Omega \log\left(\frac{dP}{dR}\right) dP & \text{if } P \ll R \\ +\infty & \text{otherwise} \end{cases}
\end{aligned}
\tag{2.1}
$$

2.2 A constraint function

One considers a function $\varphi : \Omega \mapsto \mathcal{X}$ on Ω with its values in a vector space $\mathcal{X}$ in separating duality with a vector space $\mathcal{Y}$. The space $\mathcal{X}$ is endowed with the σ-field generated by the linear forms $x \in \mathcal{X} \mapsto \langle x, y\rangle \in \mathbb{R}$, $y \in \mathcal{Y}$. It is also assumed that φ is measurable in the following sense: $\omega \in \Omega \mapsto \langle y, \varphi(\omega)\rangle \in \mathbb{R}$ is measurable for all $y \in \mathcal{Y}$.

2.3 Cramér's theorem

The sequence $X_i = \varphi(Z_i)$, $i \geq 1$ is independent identically distributed on $\mathcal{X}$ with common law $R \circ \varphi^{-1}$. A weak version of Cramér's theorem states that under the *weak Cramér condition:*

$$
\forall y \in \mathcal{Y}, \exists \lambda > 0, \int_\Omega e^{\lambda\langle y,\varphi(\omega)\rangle}\, R(d\omega) < \infty \tag{2.2}
$$

the empirical means that $1/n \sum_{i=1}^{n} \varphi(Z_i)$ obey the LDP in $\mathcal{X}$ for the topology $\sigma(\mathcal{X}, \mathcal{Y})$ with the good rate function

$$J(x) = \sup(\Delta_x), x \in \mathcal{X} \tag{2.3}$$

where (Δ_x) is the following optimization problem

$$\text{maximize } y \mapsto \langle x, y \rangle - \log \int_\Omega e^{\langle y, \varphi \rangle} \, dR, \, y \in \mathcal{Y} \tag{2.4}$$

The proof of this result is obtained using Cramér's theorem in $\mathbb{R}^d$ with the law $R \circ (\langle y_1, \varphi(\cdot)\rangle, \ldots, \langle y_d, \varphi(\cdot)\rangle)^{-1}$ ([13], Corollary 6.1.6) together with Dawson–Gärtner's theorem on the projective limits of LDPs ([13], Theorem 4.6.9).

2.4 A dual equality

In the particular case where the constraint function satisfies the following *very strong Cramér condition:*

$$\forall y \in \mathcal{Y}, \langle y, \varphi(\cdot)\rangle \in B \tag{2.5}$$

the application $P \in \mathcal{P}(\Omega) \mapsto \int_\Omega \varphi \, dP \in \mathcal{X}$, where as a definition: $\langle \int_\Omega \varphi \, dP, y \rangle_{\mathcal{X},\mathcal{Y}} = \int_\Omega \langle y, \varphi \rangle \, dP$, $\forall y \in \mathcal{Y}$, is $\sigma(\mathcal{P}(\Omega), B)$-$\sigma(\mathcal{X}, \mathcal{Y})$-continuous. It follows from the contraction principle ([13], Theorem 4.2.1) that the rate functions of Cramér's and Sanov's LDPs satisfy $J(x) = \inf\{I(P \mid R); P$, such that $\int_\Omega \varphi \, dP = x\}$, $x \in \mathcal{X}$. In other words, taking (2.3) into account, under the assumption (2.5), the following dual equality holds:

$$\inf(\Pi_x) = \sup(\Delta_x) \tag{2.6}$$

where (Π_x) is the (primal) optimization problem

$$\text{minimize } P \in \mathcal{P}(\Omega) \mapsto I(P \mid R) \text{ subject to } \int_\Omega \varphi \, dP = x \tag{2.7}$$

whose dual problem is precisely (Δ_x) (see (2.4)).

2.5 Csiszár's I-projection

As a consequence, under (2.5), one obtains the following *existence* result: If $x \in \mathcal{X}$ is such that $\sup(\Delta_x) < \infty$, there exists $P \in \mathcal{P}(\Omega)$ such that $\int_\Omega \varphi \, dP = x$ and $I(P \mid R) < \infty$ (hence $P \ll R$). One may ask what are the minimizers of (Π_x). In practice, this corresponds to a *construction* problem (see the huge literature on the entropy maximum, for instance [1] and the references therein). As $I(\cdot \mid R)$ is inf-compact (a good rate function) for the topology $\sigma(\mathcal{P}(\Omega), B)$, the minimizers are attained. On the other hand, $I(\cdot \mid R)$ is strictly convex and $\{P \in \mathcal{P}(\Omega); \int_\Omega \varphi \, dP = x\}$ is a convex set; therefore the minimizer $P_x = \text{argmin}(\Pi_x)$ is unique.

Definition 2.1. (Csiszár, [10]). *P_x is the* I-projection *of R on the convex set $\{P \in \mathcal{P}(\Omega); \int_\Omega \varphi \, dP = x\}$.*

3 Applications to stochastic processes

Clearly, condition (2.5) is very restrictive. Nevertheless, many interesting problems do not violate it. In this section, such examples are presented.

Let us take $\Omega = C([0,1], \mathbb{R}^d)$: the space of continuous paths on $[0,1]$ in $\mathbb{R}^d$. Our reference probability measure R is a process law: $R \in \mathcal{P}(\Omega)$. For any $P \in \mathcal{P}(\Omega)$ and $0 \le t \le 1$, let us denote $P_t \in \mathcal{P}(\mathbb{R}^d)$ the t-marginal of P : the law of the position at time t.

3.1 Schrödinger's bridges

Our aim is to build a process $P \in \mathcal{P}(\Omega)$ such that $P \ll R$, $P_0 = x_0$ and $P_1 = x_1$ where P_0 and P_1 are the initial and final laws of the process and x_0, x_1 are prescribed probability measures on $\mathbb{R}^d$. These constraints are properly described by the constraint function $\varphi : \omega = (\omega_t)_{0 \le t \le 1} \in \Omega \mapsto (\delta_{\omega_0}, \delta_{\omega_1}) \in \mathcal{X}$ where $\mathcal{X} = \mathcal{P}(\mathbb{R}^d) \times \mathcal{P}(\mathbb{R}^d)$ is in separating duality with $\mathcal{Y} = B(\mathbb{R}^d) \times B(\mathbb{R}^d)$ or $\mathcal{Y} = C_o^\infty(\mathbb{R}^d) \times C_o^\infty(\mathbb{R}^d)$. We denote $B(\mathbb{R}^d)$ the space of numerical bounded functions on $\mathbb{R}^d$ and $C_o^\infty(\mathbb{R}^d)$ the space of infinitely differentiable numerical functions on $\mathbb{R}^d$ with a compact support. Indeed, for any $(y_0, y_1) \in B(\mathbb{R}^d) \times B(\mathbb{R}^d)$, $\langle \int_\Omega \varphi\, dP, (y_0, y_1)\rangle_{\mathcal{X},\mathcal{Y}} = \int_\Omega [y_0(\omega_0) + y_1(\omega_1)]\, P(d\omega) = \int_{\mathbb{R}^d} y_0\, dP_0 + \int_{\mathbb{R}^d} y_1\, dP_1$. Hence, $\int_\Omega \varphi\, dP = (P_0, P_1)$. The very strong Cramér condition (2.5) clearly holds. As the dual equality (2.6) holds, an existence criterion on $x = (x_0, x_1) \in \mathcal{X}$ for such a bridge is $\sup(\Delta_x) < \infty$. This means that

$$
\begin{aligned}
&\sup_{y_0, y_1 \in B(\mathbb{R}^d)} \left\{ \int_{\mathbb{R}^d} y_0\, dx_0 + \int_{\mathbb{R}^d} y_1\, dx_1 - \log \int_\Omega e^{y_0(\omega_0) + y_1(\omega_1)}\, R(d\omega) \right\} \\
= &\sup_{y_0, y_1 \in C_o^\infty(\mathbb{R}^d)} \left\{ \int_{\mathbb{R}^d} y_0\, dx_0 + \int_{\mathbb{R}^d} y_1\, dx_1 - \log \int_\Omega e^{y_0(\omega_0) + y_1(\omega_1)}\, R(d\omega) \right\} \\
= &\sup_{y_0, y_1 \in C_o^\infty(\mathbb{R}^d)} \left\{ \int_{\mathbb{R}^d} y_0\, dx_0 + \int_{\mathbb{R}^d} y_1\, dx_1 - \log \int_{\mathbb{R}^d \times \mathbb{R}^d} e^{y_0(a) + y_1(b)}\, R_{01}(da db) \right\} \\
< &\ \infty
\end{aligned}
$$

where $R_{01}$is the joint law of (ω_0, ω_1) under R. The first equality follows from (2.3) with two different choices of $\mathcal{Y}$: $\mathcal{Y}$ has only to separate $\mathcal{X}$ and satisfy (2.5).

Note that this criterion of existence for a bridge is equivalent to the criterion of existence for a joint law P_{01} with marginal laws x_0 and x_1 such that $I(P_{01} \mid R_{01}) < \infty$.

3.2 Nelson's diffusion processes

Our aim is to build a process $P \in \mathcal{P}(\Omega)$ such that $P \ll R$, $P_t = x_t$ for all $0 \le t \le 1$ and $(x_t)_{0 \le t \le 1} \in \mathcal{X} = C([0,1], \mathcal{P}(\mathbb{R}^d))$ is a prescribed flow of t-marginals. These constraints are properly described by the constraint function $\varphi : \omega = (\omega_t)_{0 \le t \le 1} \in \Omega \mapsto (\delta_{\omega_t})_{0 \le t \le 1} \in \mathcal{X}$. The very strong Cramér condition

(2.5) clearly holds. Let us choose $\mathcal{Y} = C_o^\infty(]0,1[,\mathbb{R}^d)$ with the duality bracket $\langle x, y\rangle = \int_{[0,1]\times\mathbb{R}^d} y(t,a)\,x_t(da)dt$. The existence criterion, $\sup(\Delta_x) < \infty$, is written as follows:

$$\sup_{y\in C_o^\infty(]0,1[,\mathbb{R}^d)} \left\{ \int_{[0,1]\times\mathbb{R}^d} y(t,a)\,x_t(da)dt - \log\int_\Omega R(d\omega)\int_0^1 e^{y(t,\omega_t)}\,dt \right\} < \infty$$

To make things easier, let us take the Wiener measure with initial law R_0 for R. In [6], the following results have been proved.

Results 3.1. (Cattiaux & Léonard, [6])

1. *$J(x) < \infty$ implies that there exists $P \ll R$ such that $P_t = x_t$ for all $0 \le t \le 1$.*
2. *We have $J(x) < \infty$ if and only if there exists a vector field $b_x \in \mathcal{H}_x$ such that*

$$\int_{[0,1]\times\mathbb{R}^d} (\partial_t + b_x\cdot\nabla + \frac{\Delta}{2}) f(t,a)\,x_t(da)dt = 0, \forall f \in C_o^\infty(]0,1[\times\mathbb{R}^d)$$

   where $\mathcal{H}_x$ is the closure of $\{\nabla g;\, g \in C_o^\infty(]0,1[\times\mathbb{R}^d)\}$ in the Hilbert space $L_2([0,1]\times\mathbb{R}^d, \overline{x})$ with $\overline{x}(dtda) = x_t(da)dt$.

   Moreover, $J(x) = \frac{1}{2}\int_{[0,1]\times\mathbb{R}^d} |b_x(t,a)|^2\,x_t(da)dt$.
3. *If $J(x) < \infty$, the I-projection P_x is the unique absolutely continuous with respect to R solution to the martingale problem associated with the generator $\partial_t + b_x\cdot\nabla + \frac{\Delta}{2}$.*

- Note that the elements of $\mathcal{H}_x$ may be very irregular.
- The proof of 3.1.1 in [6] is different from the above proof.

To make precise the type of information carried by $J(x) < \infty$, we give a short proof of the necessary condition of 3.1.2.

Proof. For all $x \in C([0,1],\mathcal{P}(\mathbb{R}^d))$, we have

$$\begin{aligned} J(x) &= \sup_{y\in C_o^\infty(]0,1[,\mathbb{R}^d)} \left\{ \langle x,y\rangle - \log\int_\Omega e^{\langle y,\varphi(\omega)\rangle}\,R(d\omega) \right\} \\ &= \sup_{y\in C_o^\infty(]0,1[,\mathbb{R}^d)} \left\{ \int_{[0,1]\times\mathbb{R}^d} y\,d\overline{x} - \log\int_\Omega \exp\left(\int_0^1 y(t,\omega_t)\,dt\right) R(d\omega) \right\} \\ &\ge \sup_{f\in C_o^\infty(]0,1[,\mathbb{R}^d)} \left\{ \int_{[0,1]\times\mathbb{R}^d} -(\partial_t + \frac{\Delta}{2}) f\,d\overline{x} - \frac{1}{2}\int_{[0,1]\times\mathbb{R}^d} |\nabla f|^2\,d\overline{x} \right\}. \end{aligned}$$

For the last inequality, consider only y of the special form

$$-y_f(t,a) = (\partial_t + \frac{\Delta}{2}) f(t,a) + \frac{1}{2}|\nabla f(t,a)|^2,\ f \in C_o^\infty(]0,1[,\mathbb{R}^d)$$

and use $\int_\Omega \exp\left(\int_0^1 y(t,\omega_t)\,dt\right) R(d\omega) = 1$ (exponential martingale). Denoting $\ell_x(f) = \int_{[0,1]\times\mathbb{R}^d} -(\partial_t + \frac{\Delta}{2}) f\, d\overline{x}$, the previous inequality implies that

$$|\ell_x(f)| \leq (J(x) + \frac{1}{2})\|\nabla f\|_{2,\overline{x}},\ f \in C_o^\infty(]0,1[,\mathbb{R}^d)$$

where $\|\cdot\|_{2,\overline{x}}$ is the norm of $L_2([0,1]\times\mathbb{R}^d, \overline{x})$. It turns out that, whenever $J(x) < \infty$, $\ell_x(f)$ only depends on ∇f, that is, $\ell_x(f) = \tilde{\ell}_x(\nabla f)$, and $\tilde{\ell}_x$ is a $\|\cdot\|_{2,\overline{x}}$-continuous linear form on $\nabla C_o^\infty(]0,1[,\mathbb{R}^d)$. Finally, by the Riesz representation theorem, there exists a unique $b_x \in \mathcal{H}_x$ such that $\ell_x(f) = \int_{[0,1]\times\mathbb{R}^d} b_x(t,a)\cdot\nabla f(t,a)\, x_t(da)dt$ for all $f \in C_o^\infty(]0,1[,\mathbb{R}^d)$, which is the desired result. ∎

3.3 About the literature

The problem of Schrödinger's bridges was settled by E. Schrödinger in 1932 [26], and investigated later by several authors, in particular, S. Bernstein [2] and R. Fortet [17]. A. Beurling [3] gave a solution in the spirit of the proof of the present article. For a stimulating presentation of this problem, see H. Föllmer's lectures in Saint-Flour [16]. The above proof already appeared in the author's paper [19].

The first proof of the existence of Nelson's diffusion process is due to E. Carlen [7]. This problem has then been investigated by many authors who have given several different solutions. The reader may look to [5] and [6] for references on the subject and two distinct solutions of this problem. Among other references about Nelson's diffusions, one may read [23], [31] and [29]. In connection with Schrödinger's bridges and Nelson's diffusions, one may be interested in Berstein's processes ([30], [9]). Applications of the I-projection to Bernstein processes are given in [8].

4 Under the strong Cramér condition

In the previous sections, the dual equality (2.6) has only been proved under the very strong Cramér condition (2.5), while Cramér's theorem in $\sigma(\mathcal{X},\mathcal{Y})$ holds under a much weaker condition. In this section, Sanov's theorem is slightly extended. As a consequence, the dual equality is recovered via the contraction principle, under the intermediate condition

$$\forall y \in \mathcal{Y},\ \int_\Omega e^{\langle y,\varphi(\omega)\rangle}\, R(d\omega) < \infty \tag{4.1}$$

which is called the *strong Cramér condition*.

In the next section, a wider extension of Sanov's theorem is stated.

4.1 A slight improvement of Sanov's theorem

Let $\mathcal{M}_\tau$ be the space of measurable functions $\{f : \Omega \to \mathbb{R}; \forall\lambda > 0, \int_\Omega e^{\lambda|f|}\, dR < \infty\}$ and $\mathcal{M}_\tau^*$ its algebraic dual space. One does not identify R-almost surely equal functions in $\mathcal{M}_\tau$. Hence, any empirical measure $1/n \sum_{i=1}^n \delta_{z_i}$ belongs to $\mathcal{M}_\tau^*$ for the duality bracket $\langle 1/n \sum_{i=1}^n \delta_{z_i}, f\rangle = 1/n \sum_{i=1}^n f(z_i)$, $f \in \mathcal{M}_\tau$.

Proposition 4.1. ([21]). *The sequence $1/n \sum_{i=1}^n \delta_{Z_i}$ obeys the* LDP *in $\mathcal{P}(\Omega) \cap \mathcal{M}_\tau^*$, for the topology $\sigma(\mathcal{P}(\Omega) \cap \mathcal{M}_\tau^*, \mathcal{M}_\tau)$ with the good rate function $I(\cdot \mid R)$.*

This result is a corollary of the extended Sanov theorem stated in next the section (Theorem 5.1); see the Remark 5.2. A direct proof is as follows. For any $\vec{f} = (f_1, \ldots, f_d) \in \mathcal{M}_\tau^d$, $1/n \sum_{i=1}^n \vec{f}(Z_i)$ obeys the LDP in $\mathbb{R}^d$ with the good rate function $\zeta \in \mathbb{R}^d \mapsto \sup_{\lambda\in\mathbb{R}^d}\{\langle\lambda, \zeta\rangle - \log \int_\Omega e^{\langle\lambda, \vec{f}\rangle}\, dR\}$. Taking the projective limit of these LDPs, one obtains the LDP for L_n in $\mathcal{M}_\tau^*$ for the topology $\sigma(\mathcal{M}_\tau^*, \mathcal{M}_\tau)$ with the good rate function (compare with (2.1))

$$H_{\mathcal{M}}(\ell) = \sup_{f\in\mathcal{M}_\tau} \left\{\langle \ell, f\rangle - \log \int_\Omega e^f\, dR\right\}, \ell \in \mathcal{M}_\tau^*. \tag{4.2}$$

One concludes by checking that $H_{\mathcal{M}} = I(\cdot \mid R)$ (see [21]).

With the contraction principle, one obtains the following:

Corollary 4.2. *The dual equality (2.6) holds under the strong Cramér condition (4.1).*

We are going to state a characterization of the I-projections P_x when φ satisfies the strong Cramér condition (4.1). This result is a particular case of some results of the author ([18], Theorems 4.4 and 4.5). Let us first introduce the relevant notion of force field.

4.2 Admissible force fields

Let $\mathcal{J}$ be a totally ordered countable index set which admits a smaller element: $\flat$. We consider a family $(n) = (n^j)_{j\in\mathcal{J}}$ of measurable linear forms on $\mathcal{X}$. For any $j \in \mathcal{J}$, let us denote $T_+^j = \{\langle n^j, \varphi\rangle > 0\} \bigcap \cap_{i<j}\{\langle n^i, \varphi\rangle = 0\}$ and $T_-^j = \{\langle n^j, \varphi\rangle < 0\} \bigcap \cap_{i<j}\{\langle n^i, \varphi\rangle = 0\}$ with the convention: $\bigcap_{i<\flat}\{\langle n^i, \varphi\rangle = 0\} = \Omega$, so that $T_+^\flat = \{\langle n^\flat, \varphi\rangle > 0\}$ and $T_-^\flat = \{\langle n^\flat, \varphi\rangle < 0\}$. We define

$$\mathcal{S} = \bigcap_{j\in\mathcal{J}} \{\langle n^j, \varphi\rangle = 0\}, \quad \mathcal{T}_+ = \bigcup_{j\in\mathcal{J}} T_+^j \quad \text{and} \quad \mathcal{T}_- = \bigcup_{j\in\mathcal{J}} T_-^j.$$

Up to a R-negligible set, $\mathcal{S}$, $\mathcal{T}_+$ and $\mathcal{T}_-$ form a measurable partition of Ω.

Let us introduce notation for the force fields. Let $\overline{z}$ be a measurable linear form on $\mathcal{X}$ and $(n) = (\langle n^j, \varphi\rangle)_{j\in\mathcal{J}}$ as above. We define the application $\langle \overline{z} + \infty\cdot(n), \varphi\rangle$:

$\Omega \to [-\infty, +\infty]$, for any $\omega \in \Omega$, by

$$\langle \overline{z} + \infty \cdot (n), \varphi(\omega) \rangle = \begin{cases} +\infty & \text{if } \omega \in \mathcal{T}_+ \\ -\infty & \text{if } \omega \in \mathcal{T}_- \\ \langle \overline{z}, \varphi(\omega) \rangle & \text{if } \omega \in \mathcal{S}. \end{cases}$$

It is a measurable application. If $(n) = 0$, $\overline{z} + \infty \cdot (n) = \overline{z}$ has no infinite value.

Definition 4.3. ([18]). One says that $\overline{z} + \infty \cdot (n)$ is an *admissible force field* if:

1. $\int_{\mathcal{S}} [\langle \overline{z}, \varphi \rangle_+ e^{\langle \overline{z}, \varphi \rangle_+} + \langle \overline{z}, \varphi \rangle_-] \, dR < \infty$ and for all $\varepsilon > 0$, $K \geq 1$, $1 \leq k \leq K$, all functions f_k such that $\int_\Omega |f_k| \log(|f_k|) \, dR < \infty$ and all $g_k \in L_\infty(R)$, there exists $y \in \mathcal{Y}$ such that
$$\left| \int_\Omega \langle \overline{z} - y, \varphi \rangle [f_k \mathbf{1}_{\langle \overline{z}, \varphi \rangle \geq 0} + g_k \mathbf{1}_{\langle \overline{z}, \varphi \rangle \leq 0}] \, dR \right| \leq \varepsilon.$$
2. For all $j \in \mathcal{J}$, $\int_{\bigcap_{i<j} \{\langle n^i, \varphi \rangle = 0\}} \langle n^j, \varphi \rangle_- \, dR < \infty$.
3. $R(\mathcal{T}_+) = 0$ and $R(\mathcal{T}_-) < \infty$.

Subscripts $+$ and $-$ stand for the nonnegative and nonpositive parts of the functions.

4.3 Characterization of the I-projections

An element x of $\mathcal{X}$ is said to be an *admissible constraint* if $J(x) < \infty$. Being the domain of a convex function, the set of all admissible constraints is a convex subset of $\mathcal{X}$.

Theorem 4.4. (Characterization of the I-projections, [18]). *Let us assume that the strong Cramér condition (4.1) holds and the σ-field $\mathcal{A}$ on Ω is R-complete.*

1. *For any admissible constraint x_o, there exists an admissible force field $z_{x_o} = \overline{z}_{x_o} + \infty \cdot (n)_{x_o}$ such that*
$$x_o = \int_\Omega \varphi e^{\langle z_{x_o}, \varphi \rangle} \, dR \quad \text{and} \tag{4.3}$$
$$J(x_o) = I(e^{\langle z_{x_o}, \varphi \rangle} \cdot R \mid R) < \infty. \tag{4.4}$$
Conversely, if x_o is associated with an admissible force field z_{x_o} by formula (4.3), then (4.4) holds.
2. *If x_o is an admissible constraint, then the minimization problem (Π_{x_o}) has a unique solution P_{x_o} in $\mathcal{P}(\Omega) \cap \mathcal{M}_\tau^*$: the set of probability measures that integrates all functions $\langle y, \varphi(\cdot) \rangle$, $y \in \mathcal{Y}$. The shape of this solution is*
$$P_{x_o} = e^{\langle z_{x_o}, \varphi \rangle} \cdot R \tag{4.5}$$

where z_{x_o} is an admissible force field. Conversely, if $\bar{z} + \infty \cdot (n)$ is an admissible force field, putting $x_o = \int_\Omega \varphi e^{\langle \bar{z}+\infty\cdot(n),\varphi\rangle}\, dR$, we have $J(x_o) < \infty$, $P_{x_o} \triangleq e^{\langle \bar{z}+\infty\cdot(n),\varphi\rangle} \cdot R$ integrates all functions $\langle y, \varphi(\cdot)\rangle$, $y \in \mathcal{Y}$ and is the unique solution of (Π_{x_o}).

If x_o stands in the relative geometric interior of the effective domain of J, $z_{x_o} = \bar{z}_{x_o}$ has no infinite component. If it stands on the geometric boundary of the effective domain of J, the field of ordered collections of outward normal vectors $(n)_{x_o}$ characterizes the minimal face of the boundary on which x_o stands. Note that x_o is in the relative geometric interior of this face and $\bar{z}_{x_o}$ characterizes x_o in this face. In cases where $\Omega = C([0,1], \mathbb{R}^d)$, (4.5) is Girsanov's formula and $e^{\langle \infty\cdot(n)_{x_o},\varphi\rangle} = \mathbf{1}_S$ (R-almost surely) is the indicator function of a set of paths with finite energy.

4.4 About the literature

A characterization of the minimizer P_x in terms of the cancellation of a gradient is given in [24] and extended in ([22], Theorem 8.10) and ([28], Theorem 2). This does not lead to the exact shape of the density of the minimizer: $\frac{dP_x}{dR}$.

A necessary condition for a density to be $\frac{dP_x}{dR}$, and a sufficient condition are stated in ([10], Theorem 3.1). Except for a finite number of moment constraints, there remains a gap between these conditions to be simultaneously necessary and sufficient. Similar conditions in more general situations are obtained in ([11], Lemma 3.4) and ([22], Theorem 8.20).

For a finite number of qualified constraints, the characterization of $\frac{dP_x}{dR}$ is given in [4] and extended in [12]. Let us mention that a qualified constraint is "interior" and it follows from our results that the force field associated with P_x does not take any infinite values.

Theorem 4.4 concludes the problem of the characterization of $\frac{dP_x}{dR}$ under the strong Cramér condition (4.1) without any topological restrictions. It definitely improves the already published related results.

5 Under the weak Cramér condition

Cramér's theorem holds under the weak Cramér condition (2.2), and we have only proved Sanov's theorem under the strong Cramér condition (4.1) (see Proposition 4.1). It is interesting to ask how to extend Sanov's theorem in order to recover Cramér's theorem with the contraction principle under the weak Cramér condition. This result has been obtained in collaboration with J. Najim [21], and it is stated below. Let us begin illustrating the situation with a simple example.

5.1 Csiszár's example

This example has been studied by I. Csiszár in [11]. Take $\Omega = [0, \infty[$, $R(d\omega) = c\frac{e^{-\tilde{\lambda}\omega}}{1+\omega^3}\, d\omega$ where $\tilde{\lambda} > 0$, c is the unspecified normalizing constant and $d\omega$ is the Lebesgue measure on $[0, \infty[$. Our constraint function is $\varphi(\omega) = \omega$, $\omega \geq 0$. Note that (2.2) holds but (4.1) fails. The log-Laplace transform of R is $\Lambda(\lambda) = \log \int_{[0,\infty[} e^{\lambda\omega}\, R(d\omega)$. Its effective domain is $]-\infty, \tilde{\lambda}]$ and its left derivative at $\tilde{\lambda} : \Lambda'(\tilde{\lambda}) \triangleq \tilde{x}$ is finite. Therefore, its convex conjugate $J = \Lambda^*$ has effective domain $]0, \infty[$ and is affine with slope $\tilde{\lambda}$ on $[\tilde{x}, \infty[$. The problem (Π_x) consists of minimizing $P \mapsto I(P \mid R)$ under the constraint $\int_{[0,\infty[} \omega\, P(d\omega) = x$. In [11], it is shown that

- for any $x > 0$, the dual equality: $\inf\{I(P \mid R);\, P, \int_{[0,\infty[} \omega\, P(d\omega) = x\} = J(x)$, holds

- for any $0 < x \leq \tilde{x}$, the infimum is attained at $P_x(d\omega) = ce^{\lambda_x \omega}\, R(d\omega)$ where $\Lambda'(\lambda_x) = x$

- for any $x > \tilde{x}$, the infimum is not attained, but any minimizing sequence (P_n) (i.e., $\lim_{n\to\infty} I(P_n \mid R) = J(x)$ and $\int_{[0,\infty[} \omega\, P_n(d\omega) = x, \forall n \geq 1$), converges in the sense of total variation to $\tilde{P}(d\omega) \triangleq P_{\tilde{x}}(d\omega) = ce^{\tilde{\lambda}\omega}\, R(d\omega) = c\frac{d\omega}{1+\omega^3}$.

Csiszár introduced in [11] the notion of *generalized I-projection* to take this phenomenon into account.

5.2 The extended Sanov theorem

Let $\mathcal{L}_\tau$ stand for the space of measurable functions on Ω that admit "a finite exponential moment":

$$\mathcal{L}_\tau = \{f : \Omega \to \mathbb{R};\, \exists\lambda > 0, \int_\Omega e^{\lambda|f|}\, dR < \infty\}$$

Its algebraic dual is denoted $\mathcal{L}_\tau^*$. Proceeding as in the proof of Proposition 4.1, one obtains by projective limits of Cramér's theorem in $\mathbb{R}^d$ that $\{L_n;\, n \geq 1\}$ obeys the LDP in $\mathcal{L}_\tau^*$ for the topology $\sigma(\mathcal{L}_\tau^*, \mathcal{L}_\tau)$ and with the good rate function

$$H_{\mathcal{L}}(\ell) = \sup_{f\in\mathcal{L}_\tau} \left\{\langle \ell, f\rangle - \log \int_\Omega e^f\, dR\right\},\, \ell \in \mathcal{L}_\tau^*$$

(compare with (2.1) and(4.2)). As $\mathcal{M}_\tau \subset \mathcal{L}_\tau$, we have $H_{\mathcal{L}}(\ell) \geq H_{\mathcal{M}}(\ell') = \begin{cases} I(\ell' \mid R) & \text{if } \ell' \in \mathcal{P}(\Omega) \\ \infty & \text{otherwise} \end{cases}$ where ℓ' is the restriction of ℓ to $\mathcal{M}_\tau$.

In [21], it is proved that $H_{\mathcal{L}}(\ell) < \infty$ implies that ℓ matches with an element of the topological dual space L'_τ of the Orlicz space $L_\tau \triangleq \mathcal{L}_\tau/R$-a.s. : the factor space of $\mathcal{L}_\tau$ for the R-a.s. equality, endowed with the Luxemburg norm $\|f\|_\tau = \inf\{a > 0; \int_\Omega \tau(f/a)\,dR \leq 1\}$ where τ is the Young function $\tau(s) = e^{|s|} - |s| - 1$, $s \in \mathbb{R}$.

Any element $\ell \in L'_\tau$ is decomposed into $\ell = \ell^a + \ell^s$, the sum of its absolutely continuous part $\ell^a = \frac{d\ell^a}{dR} \cdot R$ with $\frac{d\ell^a}{dR} \in L_{\tau^*}$ and of its singular part $\ell^s \in L^s_\tau$. The space L_{τ^*} is the Orlicz space associated with the convex conjugate τ^* of τ : $\tau^*(t) = (|t| + 1)\log(|t| + 1) - |t|$, $L_{\tau^*} = \{g : \Omega \to \mathbb{R}; \int_\Omega \tau^*(g)\,dR < \infty\}$. The space L^s_τ of singular forms consists of all $\ell \in L'_\tau$ such that $\langle \ell, f\rangle = 0$, for all $f \in \mathcal{M}_\tau$. Therefore, one can write $L'_\tau \simeq (L_{\tau^*} \cdot R) \oplus L^s_\tau$.

Let $\mathcal{Q}$ be the set of all $\ell \in \mathcal{L}^*_\tau$ that are nonnegative: $\langle \ell, f\rangle \geq 0, \forall f \geq 0$, with unit mass: $\langle \ell, \mathbf{1}\rangle = 1$. Note that since $\mathbf{1}$ belongs to $\mathcal{M}_\tau$, we have $\langle \ell, \mathbf{1}\rangle = 0$ for all $\ell \in L^s_\tau$. It comes out that except for 0, the elements of L^s_τ cannot be represented as measures (they are finitely additive, but not σ-additive set functionals).

Theorem 5.1. (Extended Sanov theorem, [21]). *The sequence of random empirical measures $\{L_n; n \geq 1\}$ obeys the* LDP *in $\mathcal{Q}$ for the topology $\sigma(\mathcal{Q}, \mathcal{L}_\tau)$ with the good rate function*

$$I(\ell) = \begin{cases} I_a(\ell^a) + I_s(\ell^s) & \text{if } \ell \in L'_\tau \\ \infty & \text{otherwise} \end{cases}$$

where $\ell = \ell^a + \ell^s$, $I_a(\ell^a) = \begin{cases} I(\ell^a \mid R) & \text{if } \ell^a \in \mathcal{P}(\Omega) \\ \infty & \text{otherwise} \end{cases}$ *and* $I_s(\ell^s) = \sup\{\langle \ell^s, f\rangle; f \in \mathcal{L}_\tau, \int_\Omega e^f\,dR < \infty\}$.

Proof. See [21]. ■

Note that if $I_s(\ell^s) < \infty$, then ℓ^s is nonnegative. The function I_s is the recession function of I_a, it is also the support functional of the convex set $\{f \in \mathcal{L}_\tau; \int_\Omega e^f\,dR < \infty\}$.

Remark 5.2. One recovers Proposition 4.1 contracting the LDP of Theorem 5.1 with the application which associates with any element of $\mathcal{L}^*_\tau$ its restriction to $\mathcal{M}_\tau$. Indeed, the restriction to $\mathcal{M}_\tau$ of any singular form (in $\mathcal{L}^s_\tau$) is zero.

The equality $\langle \ell, \varphi\rangle = x$ with $\ell \in \mathcal{Q}$ and $x \in \mathcal{X}$ means $\langle \ell, \langle y, \varphi\rangle\rangle = \langle x, y\rangle$, for all $y \in \mathcal{Y}$. In view of the continuity of $\ell \in \mathcal{Q} \mapsto \langle \ell, \varphi\rangle \in \mathcal{X}$ with respect to the topologies $\sigma(\mathcal{Q}, \mathcal{L}_\tau)$ and $\sigma(\mathcal{X}, \mathcal{Y})$, the contraction principle yields the following corollary.

Corollary 5.2. *Let us consider the following extension $(\overline{\Pi}_x)$ of (Π_x) :*

$$\text{minimize } \ell \in \mathcal{Q} \cap L'_\tau \mapsto I(\ell^a \mid R) + I_s(\ell^s) \text{ subject to } \langle \ell, \varphi\rangle = x \qquad (5.1)$$

The dual equality $\sup(\Delta_x) = \inf(\overline{\Pi}_x)$ *holds under the weak Cramér condition (2.2).*

Since I is a good rate function, when $J(x)(:= \sup(\Delta_x)) < \infty$, the infimum is attained for $(\overline{\Pi}_x)$, although it may not be attained for (Π_x).

As I_s is positively homogeneous, it is not strictly convex and $(\overline{\Pi}_x)$ may admit several solutions.

Definition 5.3. The minimizers of $(\overline{\Pi}_x)$ are called *entropic projections.*

5.3 Back to Csiszár's example

Now we consider $(\overline{\Pi}_x)$ instead of (Π_x) in Csiszár's example. One can prove ([20]) that ℓ_x is a solution to $(\overline{\Pi}_x)$ if and only if

$$\ell_x = \begin{cases} P_x & \text{if } x \leq \tilde{x} \\ \tilde{P} + (x - \tilde{x})\xi & \text{if } x \geq \tilde{x} \end{cases} \quad \text{where } \xi \text{ is any nonnegative element of } L_\tau^s$$

such that

1. the constraint $\langle \xi, \varphi \rangle = 1$ is satisfied and
2. its "support" is determined by $\langle \xi, f \rangle = 0$ for all nonnegative f in L_τ such that there exists $t > 0$ with $\int_{[0,\infty[} e^{\tilde{\lambda}\varphi + tf}\, dR < \infty$.

Note that

- for all $x \leq \tilde{x}$, $\langle \ell_x, \varphi \rangle = \langle P_x, \varphi \rangle = x$ and $I(\ell_x) = I(P_x \mid R) = J(x)$ and
- for all $x \geq \tilde{x}$, $I(\ell_x) = I(\tilde{P} \mid R) + \langle (x - \tilde{x})\xi, \tilde{\lambda}\varphi \rangle = J(\tilde{x}) + \tilde{\lambda}(x - \tilde{x}) = J(x)$.

For the second item one can prove that under the constraint $\langle \xi, \varphi \rangle = 1$, the supremum of $f \mapsto \langle \xi, f \rangle$ subject to $\int_{[0,\infty[} e^f\, dR < \infty$ is attained at $\tilde{\lambda}\varphi$.

5.4 About the literature

Improvements of the usual Sanov theorem for the topology $\sigma(\mathcal{P}(\Omega), B))$ have been obtained by P. Eischelbacher and U. Schmock ([14], [15]). They are close to Proposition 4.1. A. Schied [25] shows that the relative entropy $I(\cdot \mid R)$ may not be inf-compact for topologies $\sigma(\mathcal{P}(\Omega), \mathcal{F})$ when $\mathcal{F}$ is not included in $\mathcal{M}_\tau$.

The dual equality in Corollary 5.2 is proved in [19] with convex analysis. In [20], the author characterizes the minimizers of (5.1). This extends Theorem 4.4. In [21], Theorem 5.1 is exploited to improve the Gibbs conditioning principle obtained by D. Stroock and O. Zeitouni [27] (see also [13] for a detailed presentation, and [11] for an alternate statement).

References

[1] Maximum Entropy and Bayesian Methods, *Proceedings of the 11th International Workshop on Maximum Entropy and Bayesian Methods of Statistical Analysis,* eds. C. R. Smith, G. J. Erickson and P. O. Neudorfer, Seattle, Kluwer, 1991.

[2] S. Bernstein, Sur les liaisons entre les grandeurs aléatoires *Vehr. des intern. Mathematikerkongr., Zürich 1932, Band 1.*

[3] A. Beurling, An automorphism of product measures, *Ann. Math.*, **72** (1960), 189–200.

[4] J. M. Borwein and A. S. Lewis, Partially-finite programming in L_1 and the existence of maximum entropy estimates *SIAM J. Optim.*, **3** (1993), 248–267.

[5] P. Cattiaux and C. Léonard, Minimization of the Kullback information of diffusion processes, *Ann. Inst. Henri Poincaré*, **30** (1994), 83–132.

[6] P. Cattiaux and C. Léonard, Large deviations and Nelson processes, *Forum Math.*, **7** (1995), 95–115.

[7] E. Carlen, Conservative diffusions, *Comm. Math. Phys.*, 94 (1984), 293–315.

[8] A.B. Cruzeiro, L. Wu, J.C. Zambrini, Bernstein processes associated with a Markov process. *Stochastic Analysis and Mathematical Physics*, (Santiago 98, Ed. R. Rebolledo). Trends in Math., Birkhaüser, 2000, 41–72.

[9] A.B.Cruzeiro, J.C. Zambrini, Malliavin calculus and euclidean quantum mechanics, I., *J. Funct. Anal.*, **96**:1 (1991), 62–95.

[10] I. Csiszár, I-Divergence Geometry of Probability Distributions and Minimization Problems, *Ann. Probab.*, **3**(1975), 146–158.

[11] I. Csiszár, Sanov property, generalized I-projection and a conditional limit theorem, *Ann. Probab.*, **12**(1984), 768–793.

[12] I. Csiszár, F. Gamboa and E. Gassiat, MEM pixel correlated solutions for generalized moment and interpolation problems, *IEEE Trans. Inform. Theory.*, **45**(Nov. 1999), 2253–2270.

[13] A. Dembo and O. Zeitouni, *Large Deviations Techniques and Applications, Second edition*, Springer-Verlag, 1998.

[14] P. Eischelsbacher and U. Schmock, Large and moderate deviations of products of empirical measures aud U-empirical processes in strong topologies, *preprint*, 1997.

[15] P. Eischelsbacher and U. Schmock, Exponential approximations in completely regular topological spaces and extensions of Sanov's theorem, *Stochastic Processes and their Applications*, **77** (1998), 233–251.

[16] H. Föllmer, Random Fields and Diffusion Processes, *Cours à l'Ecole d'Été de Probabilités de Saint-Flour*, Lecture Notes in Math. 1362, Springer Verlag, 1988.

[17] R. Fortet, Résolution d'un système d'équations de M. Schrödinger, *J. Math. Pures. Appl.*, **IX**(1940), 83–105.

[18] C. Léonard, Minimizers of energy functionals, to appear in *Acta Mathematica Hungarica*.

[19] C. Léonard, Minimization of energy functionals applied to some inverse problems, to appear in *Journal of Applied Mathematics and Optimization*.

[20] C. Léonard, Minimizers of energy functionals under not very integrable constraints, preprint Ecole Polytechnique, CMAP, 2000.

[21] C. Léonard and J. Najim, An extension of Sanov's theorem. Application to the Gibbs conditioning principle, preprint Ecole Polytechnique, CMAP, 2000.

[22] F. Liese and I. Vajda, *Convex Statistical Distances*, Leipzig, B.G. Teubner, 1987.

[23] E. Nelson, *Quantum Fluctuations*, Princeton Series in Physics, Princeton University Press, 1985.

[24] L. Rüschendorf, On the minimum discrimination information theorem, *Statist. Decisions*, Supplementary volume 1, 1984.

[25] A. Schied, Cramér's condition and Sanov's theorem, *Statitics and Probability Letters*, **39**(1998), 55–60.

[26] E. Schrödinger, Sur la théorie relativiste de l'électron et l'interprétation de la mécanique quantique. *Ann. Inst. Henri Poincaré*, **2**(1932), 269–310.

[27] D.W. Stroock and O. Zeitouni, Microcanonical distributions, Gibbs states, and the equivalence of ensembles, in: *Festchrift in honour of F. Spitzer*, R. Durrett and H. Kesten, eds., Birkhäuser Boston, 1991, 399–424.

[28] M. Teboulle and I. Vajda, Convergence of Best ϕ-Entropy Estimates, *IEEE Trans. Inform. Theory*, **9**(Jan. 1993), 297–301.

[29] L. Wu, Uniqueness of Nelson's diffusions, *Probab. Th. and Rel. Fields*, **114** (1999), 549–585.

[30] J.C. Zambrini, Variational processes and stochastic versions of mechanics, *J. Math. Phys.*, **9**(1986), 2307–2330.

[31] W.A. Zheng, Tightness results for laws of diffusion processes, application to stochastic mechanics, *Ann. Inst. Henri Poincaré*, **B 21**(1985), 103–124.

Modal-X, Université Paris 10, Bât. G
200, Av. de la République. 92001 Nanterre Cedex, France
Centre de Mathématiques Appliquées, Ecole Polytechnique
91128 Palaiseau Cedex, France
Christian.Leonard@u-paris10.fr

Mehler-Type Semigroups on Hilbert Spaces and Their Generators

Paul Lescot

In this paper, we shall give a survey of recent work ([1], [3], [5], [6]) on generalized Mehler semigroups by Michaël Röckner and his collaborators.[1]

1 General definitions and position of the problem

It is well-known (see the Appendix) that the Ornstein–Uhlenbeck semigroup on an abstract Wiener space (X, H, μ) can be expressed in the following way:

$$\forall t \geq 0 \;\forall f \in L^2(X) \;\forall x \in X \;\; P_t f(x) = \int_X f(e^{-t}x + \beta_t y)\mu(dy)\,, \qquad (1.1)$$

where we have set

$$\beta_t = \sqrt{1 - e^{-2t}}$$

(*Mehler formula* ([8])).

By defining:

$$\begin{aligned} T_t &: X \to X \\ x &\mapsto e^{-t}x, \\ \gamma_t &: X \to X \\ y &\mapsto \beta_t y\,, \end{aligned}$$

and

$$\mu_t = (\gamma_t)_* \mu\,,$$

we may rewrite formula (1.1) as:

$$P_t f(x) = \int_X f(T_t x + y)\mu_t(dy)\,, \qquad (1.2)$$

[1] The text is based on talks given at Bielefeld (Conference "Dirichlet Forms and Applications", August 10, 1998), Saint-Quentin (Groupe de Travail "Probabilités et Equations aux Dérivées Partielles", February 10, 1999) and Lisboa (Conference "Stochastic Analysis and Mathematical Physics", May 27, 1999). I am happy to thank, for the invitations to give these talks, Professors Michaël Röckner, Hélène Airault and Philippe Souplet, and Ana Bela Cruzeiro and Jean-Claude Zambrini, respectively.

where, clearly, $(T_t)_{t\geq 0}$ is a strongly continuous semigroup of linear operators on E, and $(\mu_t)_{t\geq 0}$ is a family of Borel probability measures on E, weakly continuous in t.

By a *generalized Mehler semigroup* on a Banach space E we shall mean a semigroup of linear operators on a function-space $\mathcal{F}(E)$ such that (1.2) hold, $(T_t)_{t\geq 0}$ and $(\mu_t)_{t\geq 0}$ then denoting respectively a given arbitrary strongly continuous semigroup of linear operators on E, and a given weakly continuous family of Borel probability measures on E. More precisely, either

(i)

$$\mathcal{F}(E) = \mathcal{C}_b(E; \mathbf{R})$$

and (1.2) holds pointwise for each $f \in \mathcal{F}(E)$; or:

(ii) $\mathcal{F}(E) = L^p(E; \nu)$ for some $p \in [1, \infty]$ and some Borel probability measure ν on E, such that (ii) makes sense and holds ν-almost surely in $x \in E$.

Obviously each such $(P_t)_{t\geq 0}$ extends naturally to $\mathcal{F}(E) \otimes_{\mathbf{R}} \mathbf{C}$ (that equals $\mathcal{C}_b(E; \mathbf{C})$ in case (i), and $L^p_{\mathbf{C}}(E; \nu)$ in case (ii)).

Let $l \in E^*$ (the topological dual of E); then

$$\begin{aligned} \forall x \in E \ \ P_t(e^{il})(x) &= \int_X e^{il(T_t x + y)} \mu_t(dy) \\ &= e^{il(T_t x)} \int_X e^{il(y)} \mu_t(dy) \\ &= \hat{\mu}_t(l) e^{i(T_t^* l)(x)}, \end{aligned} \tag{1.3}$$

where T_t^* denotes the semigroup (on E^*) *dual* to $(T_t)_{t\geq 0}$. Therefore:

$$\forall t \geq 0 \ \forall l \in E^* \ \ P_t(e^{il}) = \hat{\mu}_t(l) e^{i(T_t^* l)} . \tag{1.4}$$

This makes it easy to prove:

Lemma 1.1. ([3], Proposition 2.2) *The semigroup property for $(P_t)_{t\geq 0}$ is equivalent to:*

$$\forall s \geq 0 \ \ \forall t \geq 0 \ \ \forall l \in E^* \hat{\mu}_{s+t}(l) = \hat{\mu}_t(l) \hat{\mu}_s(T_t^* l) . \tag{1.5}$$

Proof. Let us assume the semigroup property for $(P_t)_{t\geq 0}$; as, in both cases,

$$\mathcal{C}_b(E; \mathbf{R}) \subset \mathcal{F}(E) ,$$

one has:

$$\forall f \in \mathcal{C}_b(E; \mathbf{R}) \ \ P_s(P_t f) = P_{s+t} f . \tag{1.6}$$

Obviously (1.5) therefore holds for all $f \in C_b(E; \mathbf{C})$; by taking $f = e^{il}$ ($l \in E^*$), one gets, using (1.4):

$$\begin{aligned} P_t f &= \hat{\mu}_t(l) e^{i(T_t^* l)}, \quad \text{whence:} \\ P_s(P_t f) &= \hat{\mu}_t(l) P_s(e^{i(T_t^* l)}) \\ &= \hat{\mu}_t(l)\hat{\mu}_s(T_t^* l) e^{i(T_s^*(T_t^* l))} \\ &= \hat{\mu}_t(l)\hat{\mu}_s(T_t^* l) e^{i(T_{s+t}^* l)} \end{aligned}$$

and:

$$P_{s+t} f = \hat{\mu}_{s+t}(l) e^{i(T_{s+t}^* l)}.$$

The use of (1.6) now yields condition (1.5).

Conversely, let us assume (1.5); then, reading the above argument in the reverse order, we obtain the equality $P_{s+t} f = P_s(P_t f)$ for $f = e^{il}$ ($l \in E^*$), and therefore also for $f = \cos(l)$ ($l \in E^*$) and $f = \sin(l)$ ($l \in E^*$). The equality now follows for all $f \in \mathcal{F}(E)$ by a classical monotone class argument (see [3], Proposition 2.2, p. 203). ■

Remark 1.2. In the case that $T_t = Id$ for all $t \geq 0$, the generalized Mehler semigroup is a convolution semigroup, and (1.5) expresses the multiplicativity of the Fourier transform — Lemma 1.1 therefore appears as a "twisted" version of a well-known result.

Let λ satisfy hypothesis:

(H1) $\lambda : E^* \to \mathbf{C}$ is continuous for the Sazonov topology and negative-definite, with $\lambda(0) = 0$.

Then, for each $t \geq 0$,

$$l \mapsto e^{-\int_0^t \lambda(T_u^* l) du}$$

is positive-definite and Sazonov-continuous on E^*. Therefore Bochner's Theorem ([7], Theorem 4.33, p. 50) can be applied, producing a Borel (probability) measure μ_t on E such that:

$$\forall l \in E^* \quad \hat{\mu}_t(l) = e^{-\int_0^t \lambda(T_u^* l) du}.$$

It is easy to check condition (1.5) for that family μ_t:

$$\begin{aligned} \hat{\mu}_{s+t}(l) &= e^{-\int_0^{s+t} \lambda(T_u^* l) du} \\ &= e^{-\int_0^t \lambda(T_u^* l) du} e^{-\int_t^{s+t} \lambda(T_u^* l) du} \\ &= e^{-\int_0^t \lambda(T_u^* l) du} e^{-\int_0^s \lambda(T_{v+t}^* l) dv} \\ &= e^{-\int_0^t \lambda(T_u^* l) du} e^{-\int_0^s \lambda(T_v^*(T_t^* l)) dv} \\ &= \hat{\mu}_t(l)\hat{\mu}_s(T_t^* l). \end{aligned}$$

The weak continuity of $(t \mapsto \mu_t)$ follows easily from Levy's Theorem; therefore, according to Lemma 1.1, $(P_t)_{t\geq 0}$ defined by (1.2) is a generalized Mehler semigroup. It is easy to see ([3], Lemma 2.1) that

$$P_t(\mathcal{C}_b(E;\mathbf{R})) \subset \mathcal{C}_b(E;\mathbf{R})\,;$$

in fact, for any given $f \in \mathcal{C}_b(E;\mathbf{R})$, the application $(t,x) \mapsto P_t f(x)$ is continuous on $\mathbf{R}_+ \times E$.

Remark 1.3. If $(T_t)_{t\geq 0}$ and $(\mu_t)_{t\geq 0}$ are given, the existence of λ follows from the (rather mild) hypothesis that, for all $\xi \in E^*$, $t \mapsto \hat{\mu}_t(\xi)$ is absolutely continuous on $\mathbf{R}_+$. It is then enough to set:

$$\forall \xi \in E^* \;\; \lambda(\xi) = \left[\frac{d}{dt}\hat{\mu}_t(\xi)\right]_{t=0}.$$

From now on, we shall take for E a Hilbert space, and let $J : E \to E^*$ denote the Riesz isomorphism given by:

$$\forall (e_1, e_2) \in E^2 \;\; J(e_1)(e_2) = \langle e_1, e_2 \rangle_E\,.$$

According to the infinite-dimensional version of the Levy–Khinchin formula ([7], p. 84, remark following Theorem 5.7.3), λ may be thus expressed:

$$\forall \xi \in E^* \;\; \lambda(\xi) = -i\langle \xi, b\rangle + \frac{1}{2}\langle \xi, R\xi\rangle - \int_E \left(e^{i\langle \xi, x\rangle} - 1 - \frac{i\langle \xi, x\rangle}{1+\|x\|^2}\right) M(dx) \tag{1.7}$$

where $b \in E$, $R \in \mathcal{L}(E^*; E)$ is such that $R \circ J : E \to E$ be an Hilbert–Schmidt operator, and M is a Levy measure on E (i.e. a Radon measure with $M(\{0\}) = 0$ and

$$\int_E (1 \wedge \|x\|^2) M(dx) < +\infty \;).$$

From the strong continuity of T_s, it follows that $\int_0^t T_s R T_s^* ds$ converges (for the norm topology on $\mathcal{L}(E^*, E)$); let then

$$R_t = \int_0^t T_s R T_s^* ds$$

and:

$$b_t = \int_0^t T_s b\, ds + \int_0^t \left[\int_E T_s x \left(\frac{1}{1+\|T_s x\|^2} - \frac{1}{1+\|x\|^2}\right) M(dx)\right] ds\,.$$

According to Theorem 3.1, p. 13 of [5], the following hypothesis

(H2):

(i)

$$\sup_{t>0} Tr(R_t) < +\infty\,,$$

(ii)

$$\int_0^{+\infty} \int_E (1 \wedge ||T_s x||^2) M(dx) ds < +\infty ,$$

(iii)

$$\forall x \in E \quad T_t x_{t} \rightrightarrows_{+\infty} 0$$

implies the existence of $b_\infty = \lim_{t \to +\infty} b_t$; setting

$$R_\infty = \lim_{t \to +\infty} R_t$$

and

$$M_\infty = \sup_{t>0} M_t ,$$

formula (1.7) defines a negative-definite function λ_∞ associated with b_∞, R_∞ and M_∞:

$$\begin{aligned} \forall \xi \in E^* \lambda_\infty(\xi) &= -i\langle \xi, b_\infty \rangle + \frac{1}{2}\langle \xi, R_\infty \xi \rangle \\ &\quad - \int_E \left(e^{i\langle \xi, x \rangle} - 1 - \frac{i\langle \xi, x \rangle}{1 + ||x||^2} \right) M_\infty(dx) . \end{aligned}$$

The Borel (probability) measure on E defined by:

$$\forall \xi \in E^* \quad \hat{\mu}(\xi) = e^{-\lambda_\infty(\xi)}$$

is invariant under $(P_t)_{t \geq 0}$, i.e. such that:

$$\forall t \geq 0 \; \forall f \in \mathcal{B}_b(E) \quad \int_E P_t f d\mu = \int_E f d\mu .$$

Now, yet another monotone class argument permits us to extend $(P_t)_{t \geq 0}$ to a contraction semigroup (also denoted by $(P_t)_{t \geq 0}$) on $L^2(E; \mu)$. Therefore, by the general theory of semigroups of linear operators (see for example [9], Corollary 2.5, p. 5), $(P_t)_{t \geq 0}$ possesses a densely defined generator $\mathcal{A}_0$:

$$D(\mathcal{A}_0) = \{ f \in L^2(E : \mu) \mid \lim_{t \to 0} t^{-1}(P_t f - f) \text{ exists in } L^2(E; \mu) \}$$

is dense in $L^2(E; \mu)$ and

$$\mathcal{A}_0 : D(\mathcal{A}_0) \to L^2(E; \mu)$$

is given by:

$$\forall f \in D(\mathcal{A}_0) \quad \mathcal{A}_0 f = \lim_{t \to 0} t^{-1}(P_t f - f) .$$

The data of the situation may henceafter be taken as $(T_t)_{t \geq 0}$, b, R and M satisfying (H2), such that λ satisfies (H1). We shall denote by A the generator of $(T_t)_{t \geq 0}$, and by $D(A) \subset E$ its *dense* domain. It therefore makes sense to introduce two other hypotheses:

(H3) E^* is generated by eigenvectors of A^*.

(H4) For each finite-dimensional subspace of E^* generated by eigenvectors of A^*, $\lambda|_F$ is differentiable.

2 Test functions and a heuristic argument

By a result of Courrège (see [4]), the most general possible generator for a Markov process on a finite-dimensional state space is given by a pseudo-differential operator—the *maximum principle* suffices to imply that. On an infinite-dimensional space, where no reference measure such as Lebesgue's exists, it would be very unnatural to restrict unduly the class of operators allowed. In order to motivate our definition of pseudodifferential operators, let us discuss in elementary terms, and without paying too much attention to the question of the domain, the case of state space $E = \mathbf{R}$. The topological dual E^* can then be identified with $\mathbf{R}$ in the usual way; by a differential operator is usually meant an operator of the shape:

$$\mathcal{L} = \sum_{j=0}^{n} a_j(x) \left(\frac{d}{dx}\right)^j ,$$

for continuous a_j's. Let $g : \mathbf{R} \to \mathbf{R}$, and let $f = \mathcal{F}(g)$ be its Fourier transform, defined by:

$$\forall x \in \mathbf{R}\ \mathcal{F}(g)(x) = \int_{\mathbf{R}} e^{ix\xi} g(\xi) d\xi .$$

Then:

$$\begin{aligned}
\mathcal{L}f(x) &= \sum_{j=0}^{n} a_j(x) f^{(j)}(x) \\
&= \sum_{j=0}^{n} a_j(x) \left(\frac{d}{dx}\right)^j \left(\int_{\mathbf{R}} e^{ix\xi} g(\xi) d\xi\right) \\
&= \sum_{j=0}^{n} a_j(x) \int_{\mathbf{R}} (i\xi)^j e^{ix\xi} g(\xi) d\xi \\
&= \int_{\mathbf{R}} \left(\sum_{j=0}^{n} a_j(x)(i\xi)^j\right) e^{ix\xi} g(\xi) d\xi \\
&= \int_{\mathbf{R}} p(x,\xi) e^{ix\xi} g(\xi) d\xi ,
\end{aligned}$$

where we have set:

$$p(x,\xi) = \sum_{j=0}^{n} a_j(x)(i\xi)^j .$$

Therefore, on the level of Fourier transforms, $\mathcal{L}$ corresponds to multiplication by a function $p(x, \xi)$, continuous in x and polynomial in ξ. The class of pseudodifferential operators is then defined in the same way, by relaxing the requirement that p be polynomial in ξ. On an infinite-dimensional space, "$g(\xi)d\xi$" is replaced by a complex-valued bounded Borel measure on E^*. These considerations make the definitions to follow rather natural. Let $\mathcal{M} = \mathcal{M}_b^{\mathbf{C}}(E^*)$ denote the vector space of complex-valued bounded Borel measures on E^*. For $\nu \in \mathcal{M}$, we denote by $\mathcal{F}(\nu)$ the Fourier transform defined by:

$$\mathcal{F}(\nu)(x) = \int_{E^*} e^{i\langle x,\xi\rangle}\nu(d\xi),$$

and we let $\mathcal{W} = \mathcal{F}(\mathcal{M}) \cap C_b(E; \mathbf{R})$; then one has:

Lemma 2.1.

$$\forall x \in E \;\; P_t\varphi(x) = \int_{E^*} e^{i\langle x, T_t^*\xi\rangle - \int_0^t \lambda(T_s^*\xi)ds}\nu(d\xi). \tag{2.1}$$

Proof. One has:

$$\begin{aligned}
\forall x \in E \;\; P_t\varphi(x) &= \int_E \varphi(T_t x + y)\mu_t(dy) \\
&= \int_E \left(\int_{E^*} e^{i\langle T_t x + y, \xi\rangle}\nu(d\xi)\right)\mu_t(dy) \\
&= \int_{E^*} \left(\int_E e^{i\langle T_t x + y, \xi\rangle}\mu_t(dy)\right)\nu(d\xi) \\
&= \int_{E^*} e^{i\langle T_t x, \xi\rangle}\hat{\mu}_t(\xi)\nu(d\xi) \\
&= \int_{E^*} e^{i\langle x, T_t^*\xi\rangle - \int_0^t \lambda(T_s^*\xi)ds}\nu(d\xi)
\end{aligned}$$

(the application of Fubini's Theorem being legitimate here because μ_t and ν are both *finite* measures). ■

Differentiating *formally* (2.1) at $t = 0$, we get the following heuristic expression for the generator:

$$\mathcal{A}_0\varphi(x) = \int_{E^*} (i\langle x, A^*\xi\rangle - \lambda(\xi))e^{i\langle x,\xi\rangle}\nu(d\xi). \tag{2.2}$$

Our main result consists in establishing rigorously formula (2.2) for all

$$\varphi \in \mathcal{W} = \mathcal{S}(E; \mathbf{R}).$$

By definition $\varphi \in \mathcal{W}$ if there exists $m \geq 1$, $f \in \mathcal{S}(\mathbf{R}^m; \mathbf{R})$ and

$$(\xi_1, \ldots, \xi_m) \in (E^*)^m$$

eigenvectors of A^* such that:

$$(\forall x \in E) \;\; \varphi(x) = f(\langle x, \xi_1\rangle, \ldots, \langle x, \xi_m\rangle).$$

With the above notation, let $\varphi \in \mathcal{W}$ be defined by f and the ξ_i ($A^*\xi_i = \theta_i\xi_i$); then there is $g_0 \in \mathcal{S}(\mathbf{R}^m; \mathbf{C})$ (the inverse Fourier transform of f) such that:

$$\forall \alpha \in \mathbf{R}^m \quad f(\alpha) = \int_{\mathbf{R}^m} e^{i\langle \alpha, \beta \rangle} g_0(\beta) d\beta .$$

Let $\nu_0(dv) = g_0(v)dv$,

$$\Pi_m : \mathbf{R}^m \to E^*$$

$$(v_1, \ldots, v_m) \mapsto \sum_{j=1}^{m} v_j \xi_j ,$$

and $\nu = (\Pi_m)_* \nu_0$ (carried by a *finite-dimensional* subspace of E^*). Then

$$\varphi = \mathcal{F}(\nu) \in \mathcal{F}(\mathcal{M}) \cap \mathcal{S}(E; \mathbf{R}) \subset \mathcal{F}(\mathcal{M}) \cap \mathcal{C}_b(E; \mathbf{R}) = \mathcal{W} .$$

We define $\mathcal{A}$ by:

$$\forall x \in E \ \forall \varphi \in \mathcal{W} \ \mathcal{A}\varphi(x) = \int_{E*} (i\langle x, A^*\xi \rangle - \lambda(\xi)) e^{i\langle x, \xi \rangle} \mathcal{F}^{-1}(\varphi)(d\xi) . \quad (2.3)$$

The main results of [6] are the following:

Theorem 2.2. *Under hypotheses (H1), (H2) and (H3), one has:*

(i) *$\mathcal{A}$ maps $\mathcal{W}$ into $L^2(E; \mu)$.*

(ii) *For each $\varphi \in \mathcal{W}$, $lim_{t\to 0} \frac{1}{t}(P_t\varphi - \varphi)$ exists (in $L^2(E; \mu)$) and equals $\mathcal{A}\varphi$; in other terms, $\mathcal{W} \subset D(\mathcal{A}_0)$ and $\mathcal{A}_0|_{\mathcal{W}} = \mathcal{A}$.*

Theorem 2.3. *Under the conjunction of hypotheses (H1), (H2), (H3) and (H4), $\mathcal{W}$ is a core for $\mathcal{A}_0$ (i.e., $\mathcal{W}$ is dense in $D(\mathcal{A}_0)$ for the graph norm defined by:*

$$\forall f \in D(\mathcal{A}_0) \ \ ||f||_{gr} = ||f||_{L^2(E;\mu)} + ||\mathcal{A}_0 f||_{L^2(E;\mu)} \ ).$$

Remark 2.4. By the Core Theorem, it is enough to check that $P_t(\mathcal{W}) \subset \mathcal{W}$.

Theorem 2.2 is established *via* the identity:

$$\forall t \geq 0 \ \forall \varphi \in \mathcal{W} \ \forall x \in E \ \ P_t\varphi(x) - \varphi(x) = \int_0^t P_s\mathcal{A}\varphi(x)\, ds , \quad (2.4)$$

where we write

$$P_s\mathcal{A}\varphi(x) = \int_E \mathcal{A}\varphi(T_s x + y)\mu_s(dy) ,$$

and then replace $\mathcal{A}\varphi$ by its definition (2.3). The right-hand side of (2.4) now becomes a triple integral, on $[0, t] \times E \times E^*$; unfortunately Fubini's Theorem need *not* apply here, and some delicate splitting of the integral is required.

As an example (see [5], p. 44), we may define λ by:

$$\lambda(a) = \int_E | < a, x > |^p \theta(dx)$$

where $0 < p \leq 2$ and θ is a finite symmetric measure on the unit sphere $S_E(0, 1)$; for instance, given $x_0 \in S_E(0, 1)$,

$$\theta = \frac{1}{2}(\delta_{x_0} + \delta_{-x_0})$$

is suitable. That λ is negative-definite and Sazonov-continuous, and we may take

$$T_t = e^{-t} Id ,$$

$$\hat{\mu}_t(\lambda) = e^{-\frac{1}{p}\lambda((1-e^{-pt})^{\frac{1}{p}} a)} ,$$

and

$$\hat{\mu}(a) = e^{-\frac{1}{p}\lambda(a)} .$$

Then one has:

$$P_t f(x) = \int_X f(e^{-t}x + \sqrt[p]{1 - e^{-pt}}\, y)\mu(dy),$$

a Gaussian-type situation for $p = 2$, but quite a different one otherwise.

3 Final remarks and open questions

In [2], Vladimir Bogachev, Michaël Röckner and the author have studied operators of the shape:

$$\mathcal{L}\varphi(x) = \int_{E^*} p(x, \xi) e^{i\langle x, \xi\rangle} \mathcal{F}^{-1}(\varphi)(d\xi)$$

where $p : E \times E^* \to \mathbf{R}$ is continuous in x and negative-definite in ξ, $\varphi \in \mathcal{F}(M)$ for M a suitable space of measures on E^*, E denoting here a countably conuclear space (e.g. the dual $\mathcal{S}'$ of the Schwartz space). Under reasonable conditions, we obtain the existence of a solution to the martingale problem associated with $-\mathcal{L}$. In the Hilbert space case, which is almost disjoint of that one (only when finite-dimensional is a Hilbert space conuclear), and for the very special kind of $\mathcal{L}$ $(=-\mathcal{A})$ that we have been considering, much more, i.e. the existence of an associated *process*, possibly after enlarging E, is known (see [5], Theorem 5.3, p. 23). Is there a definite common framework for all these results?

4 Appendix: The classical Mehler formula

This result is due to Mehler ([8]). Let (X, H, μ) be an abstract Wiener space, i.e. X is a Banach space, μ is a Gaussian Borel probability measure on X and H is the associated reproducing kernel Hilbert space (that equals the Cameron–Martin space in the usual case). Let $(e_n)_{n\in\mathbf{N}}$ be an orthonormal basis of H contained in X^*, and let $(H_n)_{n\in\mathbf{N}} \in \mathbf{R}[t]^{\mathbf{N}}$ denote the family of Hermite polynomials (defined by:

$$H_0(t) = 1$$

and

$$H_{n+1}(t) = tH_n(t) - H_n'(t)\,).$$

Let then D_n denote the set of functions of the type

$$x \mapsto H_{n_1}(\langle x, e_1\rangle)\ldots H_{n_k}(\langle x, e_k\rangle)$$

with $n_1 + \cdots + n_k = n$, and let C_n denote the n-th Wiener chaos, i.e. the closure of D_n in $L^2(X)$.

It is well known that

$$L^2(X;\mu) = \bigoplus_{n=0}^{+\infty} C_n\,.$$

From this follows, by a theorem from [10], the existence of a *closed* operator $\mathcal{L}$ on $L^2(X)$ such that:

$$\forall n \in \mathbf{N}\ \forall h \in C_n\ \ \mathcal{L}h = -nh\,.$$

Then the semigroup $P_t = e^{t\mathcal{L}}$ is well-defined on $L^2(E;\mu)$; as it is positive and Markovian, general theorems allow us to extend it to a contraction semigroup on $L^p(X;\mu)$ $(p > 1)$. It is given by:

Theorem 4.1.

$$P_t f(x) = \int_X f(e^{-t}x + \sqrt{1-e^{-2t}}y)\mu(dy)$$

(Mehler formula).

Proof. The set of linear combinations of functions of the type e^{il} $(l \in X^*)$ being dense in $L^2(X;\mu)$, it is enough, by a continuity argument, to verify the formula for $f = e^{il}$. For $l = 0$ it is obvious, as then $f = 1 \in \mathbf{C}_0$; we shall henceforth assume $l \neq 0$, and set:

$$a = \frac{l}{\|l\|_H}\,.$$

It is easily checked, by induction over n, that

$$H_n(x) = (-1)^n e^{\frac{x^2}{2}} \left(\frac{d}{dx}\right)^n (e^{-\frac{x^2}{2}})\,.$$

Then Taylor's formula yields:

$$e^{-\frac{1}{2}(z-y)^2} = \sum_{n=0}^{+\infty}(-1)^n e^{-\frac{z^2}{2}}(-1)^n \frac{y^n}{n!} H_n(z)\,,$$

whence the identity:

$$e^{zy-\frac{y^2}{2}} = \sum_{n=0}^{+\infty} \frac{H_n(z)}{n!} y^n\,. \tag{4.1}$$

Taking $y = ie^{-t}||l||_H$ and $z = a(x)$ in (4.1), we get:

$$e^{ie^{-t}l(x)+\frac{1}{2}e^{-2t}||l||_H^2} = \sum_{n=0}^{+\infty} \frac{H_n(a(x))}{n!}(ie^{-t}||l||_H)^n \tag{4.2}$$

and

$$\begin{aligned} \int_X f(e^{-t}x + \sqrt{1-e^{-2t}}y)\mu(dy) &= \int_X e^{il(e^{-t}x+\sqrt{1-e^{-2t}}y)}\mu(dy) \\ &= e^{ie^{-t}l(x)}\hat{\mu}(\sqrt{1-e^{-2t}}l) \\ &= e^{ie^{-t}l(x)}e^{-\frac{1}{2}(1-e^{-2t})||l||_H^2}\,, \end{aligned} \tag{4.3}$$

as μ is Gaussian. But putting $t = 0$ in (4.2) yields:

$$e^{il} = e^{-\frac{1}{2}||l||_H^2} \sum_{n=0}^{+\infty} \frac{i^n}{n!} H_n(a)\,.$$

As $H_n(a) \in \mathbf{C}_n$, one has

$$P_t(H_n(a)) = e^{-nt} H_n(a)\,.$$

Whence:

$$\begin{aligned} P_t(f)(x) &= P_t(e^{il})(x) \\ &= e^{-\frac{1}{2}||l||_H^2} \sum_{n=0}^{+\infty} \frac{i^n}{n!}||l||_H^n P_t(H_n(a))(x)) \\ &= e^{-\frac{1}{2}||l||_H^2} \sum_{n=0}^{+\infty} \frac{i^n}{n!}||l||_H^n e^{-nt} H_n(a(x)) \\ &= e^{-\frac{1}{2}||l||_H^2} e^{ie^{-t}l(x)+\frac{1}{2}e^{-2t}||l||_H^2} \quad \text{(by (4.2))} \\ &= e^{ie^{-t}l(x)}e^{-\frac{1}{2}(1-e^{-2t})||l||_H^2} \\ &= \int_X f(e^{-t}x + \sqrt{1-e^{-2t}}y)\mu(dy) \quad \text{(by (4.3))}. \end{aligned}$$

■

References

[1] V. Bogachev and M. Röckner, Mehler formula and capacities for infinite dimensional Ornstein–Uhlenbeck processes with general linear drift, *Osaka J. Math.*, **32** (1995), 237–274.

[2] V. Bogachev, P. Lescot, and M. Röckner, The martingale problem for pseudo-differential operators on infinite-dimensional spaces, *Nagoya Mathematical Journal*, **153** (1999), 101–118.

[3] V. Bogachev, M. Röckner, and B. Schmuland Generalized Mehler semigroups and applications, *Probability Theory and Related Fields*, **105** (1996), 193–225.

[4] Ph. Courrège, Sur la forme intégro-différentielle du générateur infinitésimal d'un semi–groupe de Feller sur une variété, *Séminaire de Théorie du Potentiel* (1965–1966), 48pp.

[5] M. Fuhrman and M. Röckner, Generalized Mehler semigroups—the non-Gaussian case, *Potential Analysis*, **12** (2000), 1–47.

[6] P. Lescot and M. Röckner, Generators of Mehler-type semigroups as pseudodifferential operators, preprint Bielefeld, 2000; to appear in *Infinite Dimensional Analysis and Quantum Probability.*

[7] W. Linde, *Probability in Banach Spaces-Stable and Infinitely Divisible Distributions*, John Wiley and Sons, 1986.

[8] F.G. Mehler, Ueber die Entwicklung einer Function von beliebig vielen Variablen nach Laplaceschen Functionen höheren Ordnung, *Journal de Crelle*, **66**(2), (1866), 161–176.

[9] A. Pazy, *Semigroups of Linear Operators and Applications to Partial Differential Equations*, Springer–Verlag, 1983.

[10] I.E. Segal, Tensor algebras over Hilbert spaces–I, *Trans. Amer. Math. Soc.*, **81** (1956), 106–134.

INSSET–Université de Picardie
48 Rue Raspail
02100 Saint-Quentin, France
paul.lescot@insset.u-picardie.fr

Singular Limiting Behavior in Nonlinear Stochastic Wave Equations

Michael Oberguggenberger and Francesco Russo

In this paper we study the semilinear stochastic wave equation

$$(\partial_t^2 - \Delta)u = F(u) + \dot{W} \text{ on } \mathbb{R}^{n+1}, \qquad u \mid \{t < 0\} = 0 \tag{1.1}$$

where F is globally Lipschitz, and the stochastic excitation $\dot{W}$ is a white noise on the half space $T = \mathbb{R}^n \times [0, \infty)$. The solution to the linear wave equation

$$(\partial_t^2 - \Delta)v = \dot{W} \text{ on } \mathbb{R}^{n+1}, \qquad v \mid \{t < 0\} = 0 \tag{1.2}$$

is known to be a generalized stochastic process in space dimensions $n \geq 2$, that is, its sample paths are distributions on $\mathbb{R}^{n+1}$. It is possible to construct solutions to the nonlinear wave equation (1.1) as Colombeau-type generalized stochastic processes [1, 18]. For various types of nonlinearities F, white noise calculus is applicable as well [7, 9, 11, 15]. We are concerned here with the limiting behavior of regularized solutions

$$(\partial_t^2 - \Delta)u_\varepsilon = F(u_\varepsilon) + \dot{W}_\varepsilon$$

$$(\partial_t^2 - \Delta)v_\varepsilon = \dot{W}_\varepsilon$$

obtained by smoothing white noise, as $\varepsilon \to 0$.

Our goal is to demonstrate that the approximate solutions to the nonlinear equation (1.1) converge to the solution of the linear equation (1.2) plus a deterministic term which essentially depends only on the behavior of the Fourier transform of F at zero. For example, assume that F is globally Lipschitz and has a limit L at infinity. Then, in space dimensions $n = 2, 3$,

$$\lim_{\varepsilon \to 0} u_\varepsilon = v + \frac{t^2}{2}L \text{ in } \mathcal{D}'(\mathbb{R}^{n+1}) \tag{1.3}$$

in the L^1-sense, that is,

$$\lim_{\varepsilon \to 0} E\left(\sup_{\varphi \in B} \left|\langle u_\varepsilon - v - \frac{t^2}{2}L, \varphi\rangle\right|\right) = 0$$

for every bounded subset B in the space of test functions $\mathcal{D}(\mathbb{R}^{n+1})$.

We present here two methods of proof. The first one is based on a study of the pathwise behavior of the regularized solutions v_ε to the linear wave equation as well as delta-wave estimates from nonlinear hyperbolic theory as in [5, 13, 14, 17]. This first method gives a good explanation of the observed effect: subsequences of the solutions $v_\varepsilon(x, t)$ tend to infinity almost surely, thus only the values of F near infinity are activated. However, this method gives the limiting behavior (1.3) only in probability.

The second method uses the Fourier transform of F. We single out a large class of nonlinearities for which the result (1.3) holds: namely, the Fourier transform of F should have *mass L at zero*. This, together with Gaussian properties of the free solution $v_\varepsilon(x, t)$ and the fact that its variance tends to infinity when $n \geq 2$, allows to demonstrate (1.3) in the L^1-sense. A similar argument actually gives convergence in the L^p-sense for $1 < p < \infty$. This generalizes the results of [1], where the case of nonlinearities F which are Fourier transforms of integrable measures, massless at 0, in space dimension $n = 2$ has been treated.

The observed triviality effect arises with stochastic initial value problems involving white noise as well. It occurs similarly in semilinear parabolic equations ($n \geq 2$), semilinear Schrödinger equations ($n \geq 1$) and semilinear elliptic equations on domains in $\mathbb{R}^n$, $n \geq 4$; these results will be published elsewhere. For a study of limiting solutions to semilinear parabolic equations with Wick renormalization we refer to [2]. The wave equation with white noise excitation has found quite some attention from the probabilistic side (e.g. [19]). The results presented here can also be seen from the viewpoint of delta-waves: in this spirit, stochastic analysis provides a range of intertesting, highly singular distributions whose effects as inputs in nonlinear equations can be studied.

We conclude this introduction by saying that in the last years semilinear stochastic wave equations have been intensively studied in the case of space dimension $n > 1$. Among the most significant references, we quote [3, 4, 10, 12, 16]. The driving noise in those papers is Gaussian and homogeneous but not necessarily with nuclear covariance. The equations there still allow classical function valued solutions; the covariance of the noise is situated at the boarder case not to get triviality effects.

1 White noise and the linear wave equation

Denote by $\mathcal{S}(T) = \mathcal{S}(\mathbb{R}^{n+1}) \,|\, T$ the space of rapidly decreasing smooth functions on $T = \mathbb{R}^n \times [0, \infty)$. Let $\Omega = \mathcal{S}'(T)$ with Σ the Borel σ-algebra generated by the weak topology. By the Bochner–Minlos theorem [6, §3.1], there is a unique probability measure μ on (Ω, Σ) such that

$$\int e^{i\langle\omega,\varphi\rangle}\, d\mu(\omega) = e^{-\frac{1}{2}\|\varphi\|^2_{L^2(T)}} \tag{1.4}$$

for $\varphi \in \mathcal{S}(T)$. *White noise with support in T* is the process

$$\dot{W} : \Omega \to \mathcal{D}'(\mathbb{R}^{n+1}) : \langle \dot{W}(\omega), \varphi \rangle = \langle \omega, \varphi \mid T \rangle,$$

a generalized Gaussian process with mean zero and variance

$$E(\dot{W}(\varphi)^2) = \|\varphi \mid T\|^2_{L^2(T)} \tag{1.5}$$

for $\varphi \in \mathcal{D}(\mathbb{R}^{n+1})$.

We shall make use of the following regularizations. Let $\psi \in \mathcal{D}(\mathbb{R}^{n+1})$ with $\int\int \psi(x,t)dxdt = 1$. We define ψ_ε by

$$\psi_\varepsilon(x,t) = \varepsilon^{-n-1}\, \psi(\frac{x}{\varepsilon}, \frac{t}{\varepsilon}).$$

For computational convenience, we shall assume in this paper that the support of ψ is contained in the interior of T and, in addition, that ψ is of the tensor product form

$$\psi(x,t) = \chi(x)\chi_0(t)$$

with $\chi \in \mathcal{D}(\mathbb{R}^n)$, $\chi_0 \in \mathcal{D}(0,\infty)$. Regularized white noise is defined as

$$\dot{W}_\varepsilon(\omega) = \dot{W}(\omega) * \psi_\varepsilon.$$

It belongs to $C^\infty(\mathbb{R}^{n+1})$ and has its support in T.

The next well-known proposition states existence and uniqueness of a stochastic solution to the linear wave equation (1.2). The white noise probability space Ω is as described above and fixed. The proof is included for completeness.

Proposition 1. *There is a unique generalized stochastic process $v : \Omega \to \mathcal{D}'(\mathbb{R}^{n+1})$ such that*

$$\operatorname{supp} v(\omega) \subset T$$

$$(\partial_t^2 - \Delta)v(\omega) = \dot{W}(\omega)$$

in $\mathcal{D}'(\mathbb{R}^{n+1})$ for all $\omega \in \Omega$.

Proof: Let S be the fundamental solution of the wave operator with support in the forward light cone. Then the convolution

$$v(\omega) = S * \dot{W}(\omega)$$

exists, has its support in T, and satisfies $(\partial_t^2 - \Delta)v(\omega) = \dot{W}(\omega)$ for all $\omega \in \Omega$. The measurability follows from the defining formula

$$\langle S * \dot{W}(\omega), \varphi \rangle = \langle \dot{W}(\omega), \check{S} * \varphi \rangle = \langle \omega, \check{S} * \varphi \mid T \rangle$$

where the hat denotes inflection. Uniqueness can be seen as follows. Assume $z(\omega)$ has its support in T and satisfies $(\partial_t^2 - \Delta)z(\omega) = 0$. Taking a mollifier ψ_ε as described above, we have

$$(\partial_t^2 - \Delta)(z * \psi_\varepsilon) = 0\,, \ \text{supp}\,(z * \psi_\varepsilon) \subset T\,.$$

By classical $\mathcal{C}^\infty$-theory, it follows that $z * \psi_\varepsilon \equiv 0$. But then $z = \lim_{\varepsilon \to 0} z * \psi_\varepsilon \equiv 0$ as well. □

2 Sample path estimates

In this section we shall obtain estimates on the regularized solution of the linear problem

$$\begin{aligned}(\partial_t^2 - \Delta)v_\varepsilon &= \dot{W} * \psi_\varepsilon \\ v_\varepsilon \mid \{t < 0\} &= 0\end{aligned} \tag{1.6}$$

where ψ_ε is a mollifier as described in the previous section, $\psi(x,t) = \chi(x)\chi_0(t)$ with supp $\chi_0 \subset (0, \infty)$. We first show that the variance of $v_\varepsilon(x,t)$ tends to infinity as $\varepsilon \to 0$, uniformly on strips $\mathbb{R}^n \times [t_0, t_1]$, $0 < t_0 < t_1$.

Proposition 2. *For every $t_0 > 0$ there is $\varepsilon_0 > 0$ and positive constants c_0, c_1, (depending only on the mollifier ψ) such that*

$$\frac{c_1}{\varepsilon} t \le E(v_\varepsilon(x,t)^2) \le \frac{c_0}{\varepsilon} t \quad (n = 3)$$

$$(c_1 \log \frac{1}{\varepsilon}) t \le E(v_\varepsilon(x,t)^2) \le (c_0 \log \frac{1}{\varepsilon}) t \quad (n = 2)$$

for $0 < \varepsilon \le \varepsilon_0$, $x \in \mathbb{R}^n$, $t \ge t_0$.

Proof. The regularized free solution is given by

$$v_\varepsilon(x,t) = \dot{W} * S * \psi_\varepsilon(x,t) = \langle \dot{W}, \theta\,(S * \psi_\varepsilon(x - .,\, t - .))\rangle$$

where θ is some cut-off function identically equal to one on T and vanishing as $t \to -\infty$. For fixed (x,t), the argument $\theta\,(S * \psi_\varepsilon(x - .,\, t - .))$ belongs to $\mathcal{D}(\mathbb{R}^{n+1})$. By the characterizing property (1.5) of white noise,

$$E(v_\varepsilon(x,t)^2) = \int_0^\infty \int_{\mathbb{R}^n} |S * \psi_\varepsilon(x - y, t - s)|^2\, dy ds\,.$$

Recalling that S is a smooth function of $t \in [0, \infty)$ with values in the space of compactly supported distributions $\mathcal{E}'(\mathbb{R}^n)$, replacing $x - y$ by y and $t - s$ by s, and observing the support properties of S and ψ_ε this is seen to equal

$$\int_0^t \| \int_0^s S(.,s - r) \star \psi_\varepsilon(.,r) dr \|^2_{L^2(\mathbb{R}^n)} ds$$

$$= \int_0^t \int_{\mathbb{R}^n} \Big| \int_0^s \frac{\sin(s-r)|\xi|}{|\xi|} \hat{\psi}_\varepsilon(\xi, r) dr \Big|^2 d\xi ds$$

by Parseval's equality; the latter hat and star denote Fourier transform, respectively convolution, effected in the x-variable only. Recalling that $\hat{\psi}_\varepsilon(\xi, r) = \hat{\chi}(\varepsilon\xi)\frac{1}{\varepsilon}\chi_0(\frac{r}{\varepsilon})$ finally yields

$$E(v_\varepsilon(x,t)^2) = \varepsilon^{3-n} \int_0^{t/\varepsilon} \int_{\mathbb{R}^n} \Big| \int_0^s \frac{\sin(s-r)|\xi|}{|\xi|} \chi_0(r) dr \Big|^2 |\hat{\chi}(\xi)|^2 d\xi ds \, .$$

For the remainder of the proof, we work out only the case $n = 3$, the case $n = 2$ being similar.

After introducing polar coordinates on $\mathbb{R}^3$ and denoting by M the spherical average of $|\hat{\chi}|^2$, we have

$$\begin{aligned} \frac{d}{dt} E(v_\varepsilon(x,t)^2) &= \frac{1}{\varepsilon} \int_0^\infty \Big| \int_0^{t/\varepsilon} \sin(\frac{t}{\varepsilon}\rho - r\rho)\, \chi_0(r) dr \Big|^2 M(\rho) d\rho \\ &\le \frac{1}{\varepsilon} \int_0^\infty M(\rho) d\rho \left(\int_0^\infty |\chi_0(r) dr| \right)^2 \le \frac{c_0}{\varepsilon} \, . \end{aligned}$$

Thus we have the upper bound

$$E(v_\varepsilon(x,t)^2) \le \frac{c_0}{\varepsilon} t \, .$$

To obtain the lower bound, we first observe that

$$\Big| \int_0^{t/\varepsilon} \sin(\frac{t}{\varepsilon}\rho - r\rho)\, \chi_0(r) dr \Big| = \frac{1}{2} \big| e^{i\frac{t}{\varepsilon}\rho}\, \hat{\chi}_0(\rho) - e^{-i\frac{t}{\varepsilon}\rho}\, \hat{\chi}_0(-\rho) \big|$$

for sufficiently small ε (uniformly in $t \ge t_0 > 0$) due to the fact that the support of χ_0 is a compact subset of $[0, \infty)$. This in turn equals

$$\Big| \sin(\frac{t}{\varepsilon}\rho)\, \hat{\chi}_0(\rho) + \frac{1}{2i} e^{-i\frac{t}{\varepsilon}\rho} \left(\hat{\chi}_0(\rho) - \hat{\chi}_0(-\rho) \right) \Big| \, .$$

Using the fact that $\hat{\chi}(0) = 1$, $\hat{\chi}_0(0) = 1$, we can find $\alpha > 0$ so that

$$M(\rho) \ge \frac{1}{2}, \; |\hat{\chi}_0(\rho) - 1| \le \frac{1}{10}, \; |\hat{\chi}_0(\rho) - \hat{\chi}_0(-\rho)| \le \frac{1}{5}$$

for $|\rho| \le \alpha$. Then $\frac{d}{dt} E(v_\varepsilon(x,t)^2)$ is estimated from below by

$$\frac{1}{2\varepsilon} \int_0^\alpha \Big| \sin(\frac{t}{\varepsilon}\rho) + \sin(\frac{t}{\varepsilon}\rho)\, (\hat{\chi}_0(\rho) - 1) + \frac{1}{2i} e^{-i\frac{t}{\varepsilon}\rho} \left(\hat{\chi}_0(\rho) - \hat{\chi}_0(-\rho) \right) \Big|^2 d\rho$$

$$\ge \frac{1}{2\varepsilon} \int_0^\alpha \left(|\sin \frac{t}{\varepsilon}\rho|^2 - \frac{2}{5} \right) d\rho = \frac{1}{4} \int_0^{\alpha/\varepsilon} (1 - \cos 2t\rho)\, d\rho - \frac{\alpha}{5\varepsilon} \ge \frac{\alpha}{20\varepsilon} - \frac{1}{8t_0}$$

for $t \geq t_0$. Thus, finally

$$E(v_\varepsilon(x,t)^2) \geq \left(\frac{\alpha}{20\varepsilon} - \frac{1}{8t_0}\right) t \geq \frac{c_1}{\varepsilon} t$$

for $t \geq t_0$ and sufficiently small ε. □

Remark 1. In the case $n = 1$, the variance of $v_\varepsilon(x,t)$ remains bounded. This reflects the fact that for $n = 1$, the solution v to the linear wave equation with white noise excitation is a regular stochastic process, namely a rotated two-dimensional Wiener process [19].

The divergence of the variance as $\varepsilon \to 0$ implies divergence almost surely for suitable subsequences, as is seen from the following lemma of Borel–Cantelli type.

Lemma 1. *Let* (Ω, Σ, μ), (Ξ, Σ', λ) *be probability spaces,* $V(\varepsilon, \xi, \omega)$ *be a family of random variables on* $\Xi \times \Omega$ *such that for each* $\xi \in \Xi$, $V(\varepsilon, \xi, .)$ *is Gaussian with mean zero and variance* $\sigma(\varepsilon, \xi)^2 = \int_\Omega V(\varepsilon, \xi, \omega)^2 d\mu(\omega)$. *Assume that* $\inf_{\xi\in\Xi} \sigma(\varepsilon, \xi) \to \infty$ *as* $\varepsilon \to 0$. *Then there is a subsequence* $\varepsilon_k \to 0$ *such that*

$$\lambda \times \mu \left\{(\xi, \omega) \in \Xi \times \Omega : \lim_{k\to\infty} |V(\varepsilon_k, \xi, \omega)| = \infty\right\} = 1\,.$$

In addition, every subsequence has a subsequence with this property.

Proof. Let $U(\varepsilon, \xi, \omega) = V(\varepsilon, \xi, \omega)/\sigma(\varepsilon, \xi)$. For fixed $\xi \in \Xi$ and $\varepsilon > 0$, $U(\varepsilon, \xi, .)$ is a Gaussian random variable on Ω with mean zero and variance one. By Fubini's theorem, the random variable $U(\varepsilon, ., .)$ on the product space $\Xi \times \Omega$ has the same Gaussian distribution. Choose $\varepsilon_k \to 0$ so that $\sigma(\varepsilon, \xi) \geq 2^{2k}$ for all $k \in \mathbb{N}$ and all $\xi \in \Xi$. Fix $\alpha > 0$ and consider the event

$$A_k = \left\{(\xi, \omega) \;:\; |U(\varepsilon_k, \xi, \omega)| \geq \alpha 2^k\right\}\,.$$

Using the Gaussian property of $U(\varepsilon, ., .)$ we have

$$(\lambda \times \mu)(A_k) \geq 1 - \alpha(2\pi)^{-1/2} 2^{-k+1}$$

and so

$$\lambda \times \mu \left(\bigcap_{k=1}^{\infty} A_k\right) \geq 1 - 2\alpha(2\pi)^{-1/2}\,.$$

Thus

$$\lambda \times \mu \left\{(\xi, \omega) \;:\; \lim_{k\to\infty} |V(\varepsilon_k, \xi, \omega)| = \infty\right\}$$

$$= \lambda \times \mu \left\{(\xi, \omega) \;:\; \lim_{k\to\infty} |U(\varepsilon_k, \xi, \omega)\sigma(\varepsilon, \xi)| = \infty\right\}$$

$$\geq \lambda \times \mu \left\{(\xi, \omega) \,:\, |U(\varepsilon_k, \xi, \omega)| \geq \alpha 2^k \text{ for all } k \in \mathbb{N}\right\}$$

$$= \lambda \times \mu \left(\bigcap_{k=1}^{\infty} A_k\right) \geq 1 - 2\alpha(2\pi)^{-1/2}.$$

The assertion follows by letting $\alpha \to 0$. □

Consider now the regularized solutions v_ε of the linear equation (1.6). For each fixed (x, t), they are mean zero Gaussian random variables on white noise probability space (Ω, Σ, μ). Taking any compact set $\Xi \subset \mathbb{R}^n \times (0, \infty)$ and letting λ be normalized Lebesgue measure on Ξ, we can apply Lemma 1 together with Proposition 2 to get a subsequence diverging almost surely on $\Xi \times \Omega$. Exhausting $\mathbb{R}^n \times (0, \infty)$ by compact sets, a diagonal sequence argument gives the following result:

Corollary 1. *Let v_ε be the solution to (1.6) in space dimension $n = 2$ or $n = 3$. Then there is a subsequence $\varepsilon_k \to 0$ such that μ-almost surely*

$$\lim_{k\to\infty} |v_{\varepsilon_k}(x, t)| = \infty$$

for almost all $x \in \mathbb{R}^n, t > 0$. In addition, every subsequence has a subsequence with this property. □

The corollary states, in particular, that μ-almost surely $|v_{\varepsilon_k}| \to \infty$ Lebesgue-almost everywhere on T.

3 Limits in the nonlinear equation

In this section we apply the previous results to the regularized nonlinear equation

$$(\partial_t^2 - \Delta)u_\varepsilon = F(u_\varepsilon) + \dot{W} * \psi_\varepsilon \qquad (1.7)$$
$$u_\varepsilon \mid \{t < 0\} = 0$$

where F is assumed to be globally Lipschitz, and the mollifiers ψ_ε are as in Section 1. In space dimension $n = 1, 2, 3$, equation (1.7) is easily seen to have a unique solution u_ε, a stochastic process with smooth paths. Indeed, fix $\rho > 0$ and let

$$K_\tau = \{(x, t) \in \mathbb{R}^{n+1} \,:\, 0 \leq t \leq \tau,\ |x| \leq \rho - t\}$$

be the conical region cut off at height τ with base the ball of radius ρ. The solution to the linear equation

$$(\partial_t^2 - \Delta)w = h,\ w \mid \{t < 0\} = 0$$

with smooth right hand side h is given by Kirchhoff's formula

$$w(x,t) = \frac{1}{4\pi}\int_0^t \frac{ds}{t-s}\int_{|y-x|=t-s} h(y,s)\,d\sigma(y)$$

for $n = 3$ and similarly by Poisson's and D'Alembert's formulas for $n = 2, 1$. In these dimensions, the estimate

$$\|w\|_{L^p(K_\tau)} \leq \tau \int_0^\tau \|h\|_{L^p(K_t)}\,dt \tag{1.8}$$

holds both for $p = 1$ and $p = \infty$. Rewriting (1.7) as an integral equation and employing the estimate (1.8) with $p = \infty$ in a fixed point argument yields the existence of a unique solution $u_\varepsilon(x,t,\omega)$, smooth with respect to $(x,t) \in \mathbb{R}^{n+1}$ and measurable with respect to $\omega \in \Omega$.

The pathwise estimates obtained in Section 2 allow to describe the behavior of the regularized solutions u_ε to the nonlinear equation when F has a limit at infinity, say

$$\lim_{|y|\to\infty} F(y) = L\,.$$

In this case, define the function a by $a(x,t) = \frac{t^2}{2}L$, $t \geq 0$; $a(x,t) = 0$, $t < 0$.

Theorem 1. *Let $n = 2$ or $n = 3$. Assume that F is globally Lipschitz and $\lim_{|y|\to\infty} F(y) = L$. Let u_ε be the smooth stochastic process solving problem (1.7) and let v_ε be the solution of the free equation (1.6). Then every subsequence of $\varepsilon \to 0$ has a subsequence $\varepsilon_k \to 0$ such that for all compact sets $K \subset \mathbb{R}^{n+1}$*

$$\lim_{k\to\infty} \|u_{\varepsilon_k} - v_{\varepsilon_k} - a\|_{L^1(K)} = 0$$

μ-almost surely.

Proof. Write

$$(\partial_t^2 - \Delta)(u_\varepsilon - v_\varepsilon - a)$$

$$= \int_0^1 F'(\sigma u_\varepsilon + (1-\sigma)(v_\varepsilon + a))d\sigma\ (u_\varepsilon - v_\varepsilon - a)\ +\ F(v_\varepsilon + a) - L$$

so that on any conical compact region K_τ the estimate (1.8) gives

$$\|u_\varepsilon - v_\varepsilon - a\|_{L^1(K_\tau)} \tag{1.9}$$

$$\leq \tau\|F'\|_{L^\infty(\mathbb{R})}\int_0^\tau \|u_\varepsilon - v_\varepsilon - a\|_{L^1(K_t)}\,dt\ +\ \tau^2\|F(v_\varepsilon + a) - L\|_{L^1(K_\tau)}\,.$$

By Corollary 1, for every subsequence there is a subsequence $\varepsilon_k \to 0$ such that $|v_{\varepsilon_k}(x,t,\omega)| \to \infty$ almost surely ($\omega \in \Omega$) almost everywhere ($(x,t) \in T$). For such members $\omega \in \Omega$, the bounded sequence $F(v_{\varepsilon_k} + a) - L$ converges to zero almost everywhere. Hence by Lebesgue's theorem and Gronwall's lemma the assertion follows. □

Corollary 2. *Under the assumptions of Theorem 1, let v be the solution to the free wave equation in Propostion 1. Then u_ε converges to $v + a$ with respect to the strong topology of $\mathcal{D}'(\mathbb{R}^{n+1})$, in probability as $\varepsilon \to 0$.*

Proof. Let q be one of the defining seminorms of the strong topology of $\mathcal{D}'(\mathbb{R}^{n+1})$. By Theorem 1, every subsequence of $\varepsilon \to 0$ has a subsequence $\varepsilon_k \to 0$ such that $q(u_{\varepsilon_k} - v_{\varepsilon_k} - a) \to 0$ almost surely. This is equivalent to convergence in probability. □

In case $\lim_{|y|\to\infty} F(y) = 0$, the function a vanishes and so the solutions u_ε to the nonlinear equation exhibit triviality in their behavior: they converge to the solution of the linear equation.

This pathwise study admits a rather intuitive interpretation of the triviality result: pathwise, at least for subsequences, the free solution tends to infinity, and hence the nonlinearity F is activated only near its limiting values at infinity. Thus the behavior of F on finite values has no influence on the solution. On the other hand, these arguments lead only to convergence in probability. We now present two further arguments of increasing generality that demonstrate first the triviality effect in the stronger sense of convergence in $L^1(\Omega)$ and, second, for a larger class of nonlinearities F.

Taking the expectation in (1.9) we get

$$E\left(\|u_\varepsilon - v_\varepsilon - a\|_{L^1(K_\tau)}\right) \leq \tau \|F'\|_{L^\infty(\mathbb{R})} \int_0^\tau E\left(\|u_\varepsilon - v_\varepsilon - a\|_{L^1(K_t)}\right) dt$$

$$+ \tau^2 E\left(\|F(v_\varepsilon + a) - L\|_{L^1(K_\tau)}\right).$$

The assertion

$$\lim_{\varepsilon\to 0} E\left(\|u_\varepsilon - v_\varepsilon - a\|_{L^1(K_\tau)}\right) = 0$$

is obtained, provided the last term on the right hand side tends to zero. But

$$E\left(\|F(v_\varepsilon + a) - L\|_{L^1(K_\tau)}\right)$$

$$= \iint_{K_\tau} E\left(|F(v_\varepsilon(x,t) - a(x,t)) - L|\right) dxdt$$

$$= \iint_{K_\tau} \int_{-\infty}^{\infty} |F(y - a(x,t)) - L| \frac{1}{\sqrt{2\pi}\sigma_\varepsilon} \exp(-\frac{y^2}{2\sigma_\varepsilon^2})\, dydxdt$$

where $\sigma_\varepsilon = E(v_\varepsilon(x,t)^2)$ denotes the variance of the free solution. Substituting $\sigma_\varepsilon y$ for y in the integral shows that this goes to zero provided $\lim_{|z|\to\infty} F(z) = L$. However, this Gaussian argument can be modified so that a larger class of nonlinear functions F can be treated, which we now define.

Definition 1 A distribution $H \in \mathcal{S}'(\mathbb{R})$ is said to *have mass L at zero* if

$$\lim_{\varepsilon\to 0} \langle H, \eta(\cdot/\varepsilon)\rangle = L\,, \tag{1.10}$$

for the function $\eta(y) = \exp(-y^2/2)$. If $L = 0$, H is said to be *massless at zero.*

Remark 2. For our applications, we shall be mainly concerned with functions whose Fourier transform has mass L at zero. For $F \in \mathcal{S}'(\mathbb{R})$ the Fourier transform $H = \mathcal{F}F$ has mass L at zero, i.e., satisfies (1.10), if and only if

$$\lim_{\varepsilon\to 0} \langle F, \frac{\varepsilon}{\sqrt{2\pi}}\eta(\varepsilon\,\cdot)\rangle = L\,, \tag{1.11}$$

noting that η is identical with its Fourier transform up to the multiplicative factor $\sqrt{2\pi}$.

Example 1. Let F be a continuous function such that the limit $\lim_{|y|\to\infty} = L$ exists. It follows easily from formula (11) that the Fourier transform $\mathcal{F}F$ has mass L at zero.

Example 2. Let G be a continuous function such that the limits $\lim_{x\to-\infty} = L_-$ and $\lim_{x\to+\infty} = L_+$ exist. It follows from the symmetry of η that the limit as $\varepsilon \to 0$ in (1.11) equals $(L_- + L_+)/2$. Thus $\mathcal{F}G$ is massless at zero iff $L_- = -L_+$. In particular, this holds when G vanishes at infinity.

Example 3. Let G be a periodic, sufficiently regular function with period π_0. Expanding G in its Fourier series, we see that the limit in (1.11) equals $\int_0^{\pi_0} G(y)dy$ so that $\mathcal{F}G$ is massless at zero iff G has mean zero along its period.

Example 4 If $G \in L^p(\mathbb{R})$ for some $p \in [1,2]$ or if $x^{-q}G(x) \in L^1(\mathbb{R})$, then $\mathcal{F}G$ is massless at zero (direct computation). If G is a tempered distribution such that $\mathcal{F}G = \mu$, an integrable measure with $\mu(\{0\}) = 0$, then $\mathcal{F}G$ is massless at zero.

The significance of this notion is exhibited by the following proposition.

Proposition 3. *Let $(V_1(\varepsilon), V_2(\varepsilon))$ be a mean-zero Gaussian vector, nondegenerate for all small $\varepsilon > 0$, and such that $\sigma_2^2(\varepsilon) = \operatorname{Var} V_2(\varepsilon) \to \infty$ as $\varepsilon \to 0$. Let $G : \mathbb{R} \to \mathbb{R}$ be a bounded function such that $\mathcal{F}G$ is massless at zero. Then*

$$E\,(G(V_1(\varepsilon))\,G(V_2(\varepsilon))) \to 0 \text{ as } \varepsilon \to 0\,.$$

Proof. The covariance matrix $\Sigma(\varepsilon)$ and its inverse are given by

$$\Sigma(\varepsilon) = \begin{pmatrix} \sigma_1^2 & \sigma_{12} \\ \sigma_{12} & \sigma_2^2 \end{pmatrix}, \quad \Sigma(\varepsilon)^{-1} = \begin{pmatrix} a_1^2 & a_{12} \\ a_{12} & a_2^2 \end{pmatrix},$$

Observing the relation

$$\det \Sigma^{-1} = a_1^2 a_2^2 - a_{12}^2 = \frac{a_1^2}{\sigma_2^2}$$

an easy algebraic computation gives that

$$\begin{aligned} &2\pi E\,(G(V_1)G(V_2)) \\ &= \int dx_1\, G(x_1) \int dx_2\, G(x_2) \det \Sigma^{-1/2} \exp\left(-\tfrac{1}{2}[a_1^2 x_1^2 + 2a_{12}x_1x_2 + a_2^2 x_2^2]\right) \\ &= \int dx_2 \tfrac{1}{\sigma_2} G(x_2) \exp\left(\tfrac{-x_2^2}{2\sigma_2^2}\right) a_1 \int dx_1\, G(x_1) \exp\left(-\tfrac{1}{2}[a_1 x_1 + \tfrac{a_{12}}{a_1} x_2]^2\right). \end{aligned}$$

The second integral is bounded by the L^∞-norm of G, while the first converges to zero by assumption and (1.11). □

Remark 3. Let $V_1(\varepsilon) = v_\varepsilon(x_1, t_1)$, $V_2(\varepsilon) = v_\varepsilon(x_2, t_2)$, where v_ε is the solution to the free wave equation (1.6). Then $(V_1(\varepsilon), V_2(\varepsilon))$ is nondegenerate when $|x_2 - x_1| \neq |t_2 - t_1|$. For example, in dimension $n = 3$, this follows from the fact that the covariance remains bounded, while the variances tend to infinity by Proposition 2. This in turn is seen by computing

$$\begin{aligned} &E\,(v_\varepsilon(x_1, t_1)v_\varepsilon(x_2, t_2)) \\ &\quad = \int_0^{t_1\wedge t_2} \int_{\mathbb{R}^3} S * \psi_\varepsilon(x_1 - z, t_1 - r) S * \psi_\varepsilon(x_2 - z, t_2 - r)\, dz dr \\ &\quad = \int_0^{(t_1\wedge t_2)/\varepsilon} \int_{\mathbb{R}^3} w(z, r) w(\tfrac{x_2 - x_1}{\varepsilon} + z, \tfrac{t_2 - t_1}{\varepsilon} + r)\, dz dr \end{aligned}$$

where $w(z, r)$ is the classical solution of the wave equation with right hand side ψ and zero initial data. Combining the support- and decay properties of the solution w in evaluating the latter integral shows that it remains bounded as $\varepsilon \to 0$ when $|x_2 - x_1| \neq |t_2 - t_1|$.

These considerations allow us to prove a stronger version of the result in Theorem 1:

Theorem 2. *Let $n = 2$ or $n = 3$. Assume that F is globally Lipschitz, bounded, and its Fourier transform $\mathcal{F}F$ has mass L at zero. Let u_ε be the smooth stochastic process solving problem (1.7) and let v_ε be the solution of the free equation (1.6). Then for all compact sets $K \subset \mathbb{R}^{n+1}$,*

$$\lim_{\varepsilon \to 0} E\,(\|u_\varepsilon - v_\varepsilon - a\|_{L^1(K)}) = 0.$$

Proof. For simplicity we may assume that $L = 0$, $a \equiv 0$, $F \geq 0$. On every conical compact region K_τ we have that

$$E\,(\|F(v_\varepsilon)\|_{L^1(K_\tau)}) \leq \left(E\left(\|F(v_\varepsilon)\|^2_{L^1(K_\tau)}\right)\right)^{1/2}$$

$$\leq \left(\iint_{K_T} \iint_{K_T} E\Big(F(v_\varepsilon(x_1,t_1))\, F(v_\varepsilon(x_2,t_2))\Big)\, dx_1 dt_1\, dx_2 dt_2 \right)^{1/2}.$$

We apply Proposition 3 to the Gaussian vector $(v_\varepsilon(x_1,t_1), v_\varepsilon(x_2,t_2))$, using Remark 3. Then Proposition 2 shows that the integral above tends to zero as $\varepsilon \to 0$. The discussion before Definition 1 entails the desired result. □

Corollary 3. *Under the assumptions of Theorem 2, let v be the solution to the free wave equation in Propostion 1. Then u_ε converges to $v + a$ with respect to the strong topology of $\mathcal{D}'(\mathbb{R}^{n+1})$, in $L^1(\Omega)$ as $\varepsilon \to 0$.* □

It is clear that convergence in $L^p(\Omega)$ for $1 < p < \infty$ can be proven by the same methods.

References

[1] S. ALBEVERIO, Z. HABA, F. RUSSO: *Trivial solutions for a non-linear two-space dimensional wave equation perturbed by space-time white noise.* Stochastics and Stochastics Reports **56**(1996), 127–160.

[2] S. ALBEVERIO, Z. HABA, F. RUSSO: *A two-space dimensional heat equation perturbed by a (Gaussian) white noise.* Preprint 97–18, LAGA Paris 13. To appear: Probab. Th. Rel. Fields.

[3] R. C. DALANG: *Extending the martingale measure stochastic integral with applications to spatially homogeneous SPDE's.* Electronic Journal of Probability **5**(1999).

[4] R. C. DALANG, N. FRANGOS: *The stochastic wave equation in two spatial dimensions.* Annals of Probab. **26**(1998), 187–212.

[5] T. GRAMCHEV: *Semilinear hyperbolic systems and equations with singular initial data.* Monatshefte Math. **112**(1991), 99–113.

[6] T. HIDA: *Brownian Motion.* Springer-Verlag, New York, 1980.

[7] T. HIDA, H.-H. KUO, J. POTTHOFF, L. STREIT: *White Noise. An infinite dimensional calculus.* Kluwer, Dordrecht 1993.

[8] H. HOLDEN, T. LINDSTRØM, B. ØKSENDAL, J. UBØE, T.-S. ZHANG: *The Burgers equation with a noisy force and the stochastic heat equation.* Comm. Part. Diff. Eqs. **19**(1994), 119 - 141.

[9] H. HOLDEN, B. ØKSENDAL, J. UBØE, T.-S. ZHANG: *Stochastic Partial Differential Equations.* Birkhäuser-Verlag, Basel, 1996.

[10] A. KARCZEWSKA, J. ZABCZYK: *A note on the stochastic wave equation.* Evolution Equations and their Applications in Physical and Life Sciences (G. Lumer and L. Weis, eds.), Proceedings of the 6th International Conference, Bad Herrenhalb, Marcel Dekker, New York, 2001.

[11] T. LINDSTRØM, B. ØKSENDAL, J. UBØE, T.-S. ZHANG: *Stability properties of stochastic partial differential equations.* Stoch. Anal. Appl. **13**(1995), 177–204.

[12] A. MILLET, M. SANZ-SOLÉ: *A stochastic wave equation in two space dimensions: smoothness of the law.* Annals of Probab. **27**(1999), 803–844.

[13] M. OBERGUGGENBERGER: *Weak limits of solutions to semilinear hyperbolic systems.* Math. Ann. **274**(1986), 599–607.

[14] M. OBERGUGGENBERGER, Y.-G. WANG: *Delta-waves for semilinear hyperbolic Cauchy problems.* Math. Nachr. **166**(1994), 317–327.

[15] B. ØKSENDAL: *Stochastic partial differential equations: a mathematical connection between macrocosmos and microcosmos.* In: M. GYLLENBERG, L.-E. PERSSON (EDS.), *Analysis, Algebra and Computers in Mathematical Research.* Lecture Notes Pure Appl. Math. Vol. 156, M. Dekker, New York 1994, 365–385.

[16] S. PESZAT, J. ZABCZYK: *Nonlinear stochastic wave and heat equations.* Probab. Th. Rel. Fields **116** (2000), 421-443.

[17] J. RAUCH AND M. REED: *Nonlinear superposition and absorption of delta waves in one space dimension.* J. Funct. Anal. **73**(1987), 152–178.

[18] F. RUSSO: *Colombeau generalized functions and stochastic analysis.* In: A.I. CARDOSO, M. DE FARIA, J. POTTHOFF, R. SÉNÉOR, L. STREIT (EDS.), *Analysis and Applications in Physics.* Kluwer, Dordrecht 1994, 329–349.

[19] J.B. WALSH: *An Introduction to Stochastic Partial Differential Equations.* In: R. CARMONA, H. KESTEN, J.B. WALSH (EDS.), *École d'Été de Probabilités de Saint Flour XIV - 1994.* Springer Lecture Notes Math. Vol. 1180, Springer-Verlag, New York 1986, 265–439.

Michael Oberguggenberger
Institut für Mathematik und Geometrie
Universität Innsbruck
A-6020 Innsbruck, Austria
email: michael@mat1.uibk.ac.at
and
Francesco Russo
Institut Galilée, Département de Mathématiques
Université Paris 13
F - 93430 Villetaneuse, France
e-mail :russo@math.univ-paris13.fr
frusso@physik.uni-bielefeld.de

Complete Positivity and Open Quantum Systems

Rolando Rebolledo*

ABSTRACT The interpretation provided by Schrödinger of quantum mechanics through classical stochastic processes is the root of Euclidean quantum mechanics founded on Bernstein processes. Open quantum systems have been extensively investigated during the last three decades and have motivated new probabilistic interpretations for quantum dynamics. This article, mainly addressed to classical probabilists, gives a short introduction to quantum dynamical semigroups and quantum Markov flows as a natural extension of Schrödinger ideas in noncommutative stochastic analysis.

This article illustrates the concept of complete positivity connected with stochastic models for open quantum systems, an idea going back to Schrödinger. We assert that quantum Markov processes, characterized through completely positive maps, translate well the original probabilistic interpretation of Schrödinger's equations in quantum mechanics.

The organization of the article is as follows. The ideas of Schrödinger are used in Section 1 to motivate the algebraic study of complete positivity. In Section 2 the definition of completely positive maps is given together with their main properties, in particular, the representation theorem due to Stinespring. Quantum dynamical semigroups and quantum Markov flows are introduced in Section 3, illustrated by several examples making use of classical stochastic noise. Quantum noise and noncommutative stochastic differential equations appear in Section 4, a section devoted to the study of an important example of an open quantum system, a class of models appearing in quantum optics. Finally, Section 5 shows how a given quantum dynamical semigroup produces numerous commutative Markov semigroups when restricted to some particular abelian algebras.

*This research has been partially supported by the Cátedra Presidencial en Análisis Cualitativo de Sistemas Dinámicos Cuánticos, FONDECYT grant 1990439 and the ICCTI/CONICYT-98 Portuguese/Chilean exchange program

1 Starting from Schrödinger bridges

Take a complex separable Hilbert space $\mathcal{H}$ and a selfadjoint operator H. The well-known Schrödinger equation is usually written (with the convention $\hbar = 1$), as

$$\frac{\partial}{\partial t}\psi_t = -iH\psi_t, \ \psi_0 = \psi \in \mathcal{H}. \tag{1.1}$$

where $\psi_t \in \mathcal{H}$ is the wave function, satisfying $\|\psi_t\| = 1$.

Since the formulation of this fundamental equation several probabilistic interpretations have been proposed. In particular, Schrödinger himself introduced a class of (classical) stochastic processes, now called *Schrödinger bridges*. These processes were later formalized by Bernstein and extensively developed by Zambrini and other authors (see [5]. On the other hand, a new stochastic analysis of a noncommutative nature, also inspired by quantum mechanics, is currently being developed, but this analysis is most influenced by the study of the *open quantum systems*. The aim of this article is to go beyond Schrödinger bridges via the concept of *complete positivity*, thus establishing contact with the rising field of quantum dynamical semigroups and quantum Markov flows.

The solution of (1.1) is given by a unitary group $U(t) = e^{-itH}$ acting on $\mathcal{H}$, that is, $\psi_t = U(t)\psi$. Let us denote $(e_n)_{n\in\mathbb{N}}$ as an orthonormal basis of $\mathcal{H}$. We use Dirac notation $|u\rangle\langle v|w = \langle v, w\rangle u$ for projectors ($u, v, w \in \mathcal{H}$). A simple computation yields

$$\begin{aligned} 1 &= \|\psi_t\|^2 \\ &= \langle\psi_t, \psi_t\rangle \\ &= \operatorname{tr} |\psi_t\rangle\langle\psi_t| \\ &= \sum_{n\in\mathbb{N}}\langle e_n, |\psi_t\rangle\langle\psi_t|e_n\rangle \\ &= \sum_{n\in\mathbb{N}}\langle e_n, U(t)|\psi\rangle\langle\psi|U(t)^*e_n\rangle \end{aligned}$$

Therefore

$$0 = d\langle\psi_t, \psi_t\rangle = \langle d\psi_t, \psi_t\rangle + \langle\psi_t, d\psi_t\rangle. \tag{1.2}$$

Formally we have

$$0 = d(U(t)|\psi\rangle\langle\psi|U(t)^*) = dU(t)|\psi\rangle\langle\psi|U(t)^* + U(t)|\psi\rangle\langle\psi|dU(t)^*.$$

For all $t \geq 0$, $\rho_t = |\psi_t\rangle\langle\psi_t|$ is a *pure state*, that is, an extremal point of the convex set of all trace class endomorphisms ρ of $\mathcal{H}$ with unit trace. The Schrödinger equation for states is then written as

$$\frac{d\rho_t}{dt} = -i[H, \rho_t], \tag{1.3}$$

where $[\cdot, \cdot]$ is the usual notation for the commutator of two operators.

Going back to Schrödinger ideas in [49], we take $\mathcal{H} = L^2(\mathbb{R}^d; \mathbb{C})$ with the usual scalar product. We consider functions f defined on the positive real line with values in $\mathcal{H}$ and define the time reversal operator $\mathfrak{r}_t$ as follows:

$$\mathfrak{r}_t f(s) = f(t-s)1_{[0,t]}(s) + f(s)1_{]t,\infty[}(s), \quad (s, t \geq 0).$$

Now we fix the time interval as $[0, 1]$ and consider the Schrödinger equation on that interval as a two-fold problem: first, $t \mapsto \psi_t = U(t)\psi_0$ solves an initial value (or *forward*) problem; second, $t \mapsto \bar{\psi}_t = (\mathfrak{r}_1 U(\cdot)\bar{\psi}_1)(t)$ is a solution to a *final* value (or *backward*) problem and $\psi_t(x)\bar{\psi}_t(x)dx$ is a probability measure on $\mathbb{R}^d$. Indeed,

$$\bar{\psi}_t = e^{-itH}\bar{\psi}_0 = e^{-i(1-t)H}\bar{\psi}_1,$$

since $\psi_1 = e^{iH}\psi_0$.

These ideas inspired Bernstein to introduce a particular class of stochastic processes (see [23], [50], [11]), connected with the above interpretation. Instead of the unitary family $U(t) = e^{-itH}$, one considers a couple of semigroups $V(t) = e^{-tH}$ and the related time reversed family $\tilde{V}(t) = \mathfrak{r}_1 V(t) = e^{-(1-t)H}$, $(t \in [0, 1])$. Given positive functions $\eta_0, \eta_1 \in \mathcal{H}$, one considers $\eta_t = V(t)\eta_0$, $\tilde{\eta}_t = \tilde{V}(t)\eta_1$ and one looks for conditions under which $\eta_t(x)\tilde{\eta}_t(x)dx$ is a probability measure on $\mathbb{R}^d$, $(t \in [0, 1])$. $U(\tau)$ is related to $V(t)$ through analytic continuation and change of variables $\tau = -it$.

The probabilistic interpretation proposed by Euclidean quantum mechanics is founded on $(V, \tilde{V})$ to which a classical stochastic process can be associated, namely, the class of Bernstein processes. For a detailed treatment of this question we refer the reader to [50], [5], [11]. Although Bernstein processes have been built up within Kolmogorov's probability model which is fundamentally commutative, the guidelines of their foundations are deeply *non commutative*. This provides a foundation for an extension of the above interpretation within a noncommutative probabilistic framework. To make this clear we would like to study more closely the algebraic properties of expressions like $U(t)\rho U^*(t)$ and $V(t)\rho\tilde{V}(t)$.

2 A primer on complete positivity

We start by extending the classical notion of a *transition kernel* in probability theory. Let two measurable spaces $(E_i, \mathcal{E}_i)$, $(i = a, b)$ and a kernel $P(x, dy)$ from E_b to E_a be given. That is, $P : E_b \times \mathcal{E}_a$ is such that

- $x \mapsto P(x, A)$ is measurable from E_b in $[0, 1]$ for any $A \in \mathcal{E}_a$;
- $A \mapsto P(x, A)$ is a probability on $(E_a, \mathcal{E}_a)$ for all $x \in E_b$.

We denote $\mathcal{A}$ (respectively $\mathcal{B}$) the algebra of all complex bounded measurable functions defined on E_a (resp. E_b). These are *-algebras (they have an involution given by the operation of complex conjugation) with unit. Moreover, they are C^*-algebras since they are complete for the topology defined by the uniform norm. The kernel P defines a linear map Φ_P from $\mathcal{A}$ to $\mathcal{B}$ given by $\Phi_P(a) = Pa$, where

$$Pa(x) = \int_{E_b} P(x, dy)a(y),$$

for all $a \in \mathcal{A}$, $x \in E_b$.

It is worth noting that Φ_P is a positive map, and that it satisfies a stronger property: for any finite collection of elements $a_i \in \mathcal{A}$, $b_i \in \mathcal{B}$, $(i = 1, \ldots, n)$, the function

$$\sum_{i,j=1}^{n} b_i \Phi_P(a_i \bar{a}_j)\bar{b}_j, \tag{2.1}$$

is positive. Indeed, for any fixed $x \in E_b$, $P(x, \cdot)$ is positive definite, so for any collection $\alpha_1, \ldots, \alpha_n$ of complex numbers, the sum

$$\sum_{i,j} \alpha_i \bar{\alpha}_j P(a_i \bar{a}_j)(x),$$

is positive. It is enough to choose $\alpha_i = b_i(x)$, $(i = 1, \ldots, n)$, to obtain (2.1).

Now take a probability measure μ on $(E_b, \mathcal{E}_b)$, call $\Omega = E_b \times E_a$, $\mathcal{F} = \mathcal{E}_b \otimes \mathcal{E}_a$ and define a probability $\mathbb{P}$ on $(\Omega, \mathcal{F})$ given by

$$\mathbb{E}(b \otimes a) = \int_{E_b} \mu(dx)b(x)Pa(x), \tag{2.2}$$

where $a \in \mathcal{A}$, $b \in \mathcal{B}$.

Under the probability $\mathbb{P}$, the random variables (X_b, X_a), given by the coordinate maps on $E_b \times E_a$, satisfy the following property: μ is the distribution of X_b and $P(x, dy)$ is the conditional probability of X_a given that $X_b = x$.

We now study the construction of $L^2(\Omega, \mathcal{F}, \mathbb{P})$. Consider the family of random variables of the form $X = \sum_{i=1}^{n} b_i \otimes a_i$ with $a_i \in \mathcal{A}$, $b_i \in \mathcal{B}$, $(i = 1, \ldots, n)$. The scalar product of two of such elements is

$$\langle X^{(1)}, X^{(2)} \rangle = \int_{E_b} \mu(dx) \sum_{i,j} b_i^{(1)}(x) P(a_i^{(1)} \bar{a}_j^{(2)})(x) \bar{b}_j^{(2)}(x). \tag{2.3}$$

Note that (2.1) is needed if one wants to define the scalar product via (2.3). Within this commutative framework, property (2.1) is granted by the positivity of the kernel. This fails in the noncommutative case.

Definition 2.1. Let two C^*-algebras $\mathcal{A}$ and $\mathcal{B}$ be given. A linear map $\Phi : \mathcal{A} \to \mathcal{B}$ is *completely positive* if for any two finite collections $a_1, \ldots, a_n \in \mathcal{A}$ and $b_1, \ldots, b_n \in \mathcal{B}$, the element

$$\sum_{i,j=1}^{n} b_i^* \Phi(a_i^* a_j) b_j \in \mathcal{B},$$

is positive.

We recall that a *representation of a C^*–algebra $\mathcal{A}$* is $(\pi, \mathcal{K})$, where $\mathcal{K}$ is a complex Hilbert space and π is a *-homomorphism of $\mathcal{A}$ and the algebra of all bounded linear operators on $\mathcal{K}$, $\mathcal{B}(\mathcal{K})$.

Two complementary results, one due to Arveson and the second proved by Stinespring, show that complete positivity is always derived from positivity in the commutative case. More precisely, we have the following.

Theorem 2.2. *A positive map $\Phi : \mathcal{A} \to \mathcal{B}$ is completely positive if at least one of the following two conditions is satisfied:*

(a) $\mathcal{A}$ is commutative (Stinespring [47]);

(b) $\mathcal{B}$ is commutative (Arveson [6]).

Moreover, the sum of completely positive maps is completely positive as well as the composition of two of such maps. Furthermore, any *-homomorphism of algebras is completely positive. Thus, given any representation $(\pi, \mathcal{K})$ of the C^*-algebra $\mathcal{A}$, π is completely positive.

Assume that $\mathcal{A}$ and $\mathcal{B}$ are C^*-algebras, $\mathcal{B} \subset \mathcal{B}(\mathcal{H})$, where $\mathcal{H}$ is a complex separable Hilbert space. Let a completely positive map $\Phi : \mathcal{A} \to \mathcal{B}$ be given. Take two arbitrary elements $x = \sum_i a_i \otimes u_i$, $y = \sum_j b_j \otimes v_j$ in $\mathcal{A} \times \mathcal{H}$, where both sums contain a finite number of terms, and define

$$\langle x, y \rangle = \sum_{i,j} \langle u_i, \Phi(a_i^* b_j) v_j \rangle.$$

Since Φ is completely positive, $\langle x, x \rangle \geq 0$. Denote

$$\mathcal{N} = \{x \in \mathcal{A} \otimes \mathcal{H};\ \langle x, x \rangle = 0\},$$

and introduce on the quotient space $(\mathcal{A} \otimes \mathcal{H})/\mathcal{N}$ the scalar product

$$\langle x + \mathcal{N}, y + \mathcal{N} \rangle = \langle x, y \rangle.$$

By completion, we obtain a Hilbert space denoted by $\mathcal{K}$.

Our purpose now is to define a *-homomorphism $\pi : \mathcal{A} \to \mathcal{B}(\mathcal{K})$. This is done in two steps. First, define $\pi_0(a)$ for $a \in \mathcal{A}$ on elements of the form x:

$$\pi_0(a) \left(\sum_i a_i \otimes u_i \right) = \sum_i (a a_i) \otimes u_i.$$

For x and y as before, $\pi_0(a)$ is a linear application in $\mathcal{A} \otimes \mathcal{H}$ that satisfies

$$\langle x, \pi_0(a) y \rangle = \langle \pi_0(a^*) x, y \rangle \tag{2.4}$$

$$\begin{aligned} \|\pi_0(a)x\|^2 = \langle x, \pi_0(a^* a) x \rangle &\leq \|a^* a\| \langle x, \pi_0(\mathbf{1}) x \rangle \\ &\leq \|a\|^2 \|x\|^2. \end{aligned} \tag{2.5}$$

From the above relations, π_0 extends into a $*$-homomorphism $\pi : \mathcal{A} \to \mathcal{B}(\mathcal{K})$ and $(\pi, \mathcal{K})$ is a representation of $\mathcal{A}$. Moreover, we can define a linear operator $V : \mathcal{H} \to \mathcal{K}$ by

$$Vu = \mathbf{1} \otimes u + \mathcal{N}.$$

This is a bounded operator since

$$\|Vu\|^2 = \langle u, \Phi(\mathbf{1})u\rangle \leq \|\Phi(\mathbf{1})\| \|u\|^2.$$

Finally, Φ may be written in the form

$$\Phi(a) = V^* \pi(a) V, \tag{2.6}$$

for all $a \in \mathcal{A}$.

On the other hand, if Φ is given by (2.6), an elementary computation shows that Φ is completely positive. Thus we have obtained the celebrated characterization of completely positive maps due to Stinespring [47] (see also [34], [39]). We assume that the C^*-algebras $\mathcal{A}$ and $\mathcal{B}$ have a unit denoted in both cases by the same symbol $\mathbf{1}$. This assumption implies in particular that our algebras are *simple* and they have faithful representations.

Theorem 2.3. [Stinespring] *Let $\mathcal{B}$ be a sub C^*-algebra of the algebra of all bounded operators on a given complex separable Hilbert space $\mathcal{H}$. Assume $\mathcal{A}$ to be a C^*-algebra with unit. A linear map $\Phi : \mathcal{A} \to \mathcal{B}$ is completely positive if and only if it has the form*

$$\Phi(x) = V^* \pi(x) V,$$

where $(\pi, \mathcal{K})$ is a representation of $\mathcal{A}$ on some Hilbert space $\mathcal{K}$, and V_Φ is a bounded operator from $\mathcal{H} \to \mathcal{K}$.

The representation (2.6) is not unique. We call (π, V) a *Stinespring representation* of Φ. Moreover, the above representation is said to be *minimal* if $\{\pi(x)Vu : x \in \mathcal{A},\ u \in \mathcal{H}\}$ is dense in $\mathcal{K}$. For a completely positive map, the minimal representation is unique up to unitary equivalence.

If the completely positive map Φ is σ-weakly continuous and preserves identity, then its minimal representation (π, V) is such that π is σ-weakly continuous and V is an isometry, $V^*V = \mathbf{1}$. We denote by $\mathbf{CP}(\mathcal{A}, \mathcal{B})$ the set of all σ-weakly continuous completely positive maps $\Phi : \mathcal{A} \to \mathcal{B}$ which preserve identity.

For a von Neumann algebra $\mathcal{A}$, and $\mathcal{B} = \mathcal{B}(\mathcal{K})$, Kraus (see [30]) obtained the following characterization of normal completely positive maps.

Theorem 2.4. [Kraus] *Let two complex separable Hilbert spaces $\mathcal{H}$, $\mathcal{K}$, and a von Neumann algebra $\mathcal{A}$ of operators of $\mathcal{H}$ be given. Then a linear map $\Phi : \mathcal{A} \to \mathcal{B}(\mathcal{K})$ is normal and completely positive if and only if there exists a sequence $(V_j)_{j\in\mathbb{N}}$ of linear bounded operators from $\mathcal{K}$ to $\mathcal{H}$ such that the series $\sum_{j=1}^{\infty} V_j^* a V_j$ strongly converges for any $a \in \mathcal{A}$ and*

$$\Phi(a) = \sum_{j=1}^{\infty} V_j^* a V_j. \tag{2.7}$$

This representation can be improved by introducing an additional arbitrary complex and separable Hilbert space $\tilde{\mathcal{H}}$ with an orthonormal basis $(f_n)_{n\in\mathbb{N}}$. Indeed, defining $V : \mathcal{K} \to \mathcal{H} \otimes \tilde{\mathcal{H}}$ by

$$Vu = \sum_j V_j u \otimes f_j, \quad (u \in \mathcal{K}),$$

we then have that

$$\Phi(a) = V^*(a \otimes \mathbf{1})V, \tag{2.8}$$

where $\mathbf{1}$ is the identity operator of $\tilde{\mathcal{H}}$, $a \in \mathcal{A}$.

The following corollary is a straightforward consequence of Stinespring's theorem.

Corollary 2.5. *Let H denote a selfadjoint bounded operator defined on a given complex Hilbert space h. Then, for any given $z \in \mathbb{C}$ the map $\Phi_z(X) = e^{-zH} X e^{zH}$, $X \in \mathcal{B}(h)$, is completely positive if and only if $\Re z = 0$.*

Proof. By Stinespring's theorem Φ_z is completely positive if and only if $e^{\bar{z}H} = (e^{zH})^* = e^{-zH}$. □

Coming back to applications such as $\rho \mapsto U(t)\rho U^*(t)$ and $\rho \mapsto V(t)\rho\tilde{V}(t)$ referred to at the end of the first section, there is a fundamental difference between them: while the first map is completely positive the second is not. However, the analytic extension of $\rho \mapsto V(z)\rho\tilde{V}(z)$ is completely positive when $z = it$, which is obvious since in this case $V(it)$ coincides with $U(t)$.

To summarize, applications of the form $\rho \mapsto V(t)\rho\tilde{V}(t)$ are not associated to a Markov structure in general, since complete positivity is a fundamental property of Markov semigroups, which we will see in the next section.

3 Quantum dynamical semigroups and Markov flows

A homogeneous classical Markov semigroup is characterized by a family $(P_t)_{t\geq 0}$ of Markovian transition kernels defined on a measurable space $(E, \mathcal{E})$ which satisfies the Chapman–Kolmogorov equations (or the semigroup property for the composition of kernels). Given a σ-finite measure μ on $(E, \mathcal{E})$, $\mathcal{A} = L^\infty(E, \mathcal{E}, \mu)$ represents the von Neumman algebra of multiplication operators acting on the Hilbert space $L^2(E, \mathcal{E}, \mu)$. In this case, the predual algebra is $\mathcal{A}_* = L^1(E, \mathcal{E}, \mu)$. Moreover, $(P_t)_{t\geq 0}$ is a semigroup of completely positive maps acting on the von Neumann algebra $\mathcal{A}$. Additionally, this semigroup satisfies the following properties:

- It preserves the unit: $P_t\mathbf{1} = \mathbf{1}$, for all $t \geq 0$.

- $P_0 = I$, the identity mapping.
- Each P_t is σ-weak continuous, that is, for any increasing net f_α of positive elements with upper envelope f in $\mathcal{A}$, we have that $\int_E P_t f(x)g(x)\mu(dx) = \lim_\alpha \int_E P_t f_\alpha(x)g(x)\mu(dx)$, for all $g \in L^1(E, \mathcal{E}, \mu)$. Indeed, by the Monotone Convergence Theorem, $P_t f_\alpha(x) \uparrow P_t f(x)$, for all $x \in E$; and to conclude, it is enough to apply the Dominated Convergence Theorem to $P_t f_\alpha(x)g(x)$.

All the above properties are crucial in order to extend Markovian concepts to a noncommutative framework. Moreover, it is well-known that the addition of suitable topological hypotheses on the space $(E, \mathcal{E})$ allows us to construct a Markov process associated to a semigroup. One can take, for instance, E to be a locally compact space with countable basis and $\mathcal{E}$ its Borel σ–field. This leads to the Markovian system

$$(\Omega, \mathcal{F}, (\mathcal{F}_t)_{t\geq 0}, (\mathbb{P}_x)_{x\in E}, (X_t)_{t\geq 0}, E, \mathcal{E}).$$

The semigroup and the process are related by the equation

$$P_t f(x) = \mathbb{E}_x(f(X_t)),$$

for all $f \in \mathcal{A}$, $t \geq 0$. Also we can choose an arbitrary initial probability ν on $(E, \mathcal{E})$, and denote $\mathbb{P}_\nu = \int \mathbb{P}_x \nu(dx)$.

Now consider the von Neumann algebra $\mathcal{B} = L^\infty(\Omega, \mathcal{F}, \mathbb{P}_\nu)$. The Markov flow is defined as a *-homomorphism $j_t : \mathcal{A} \to \mathcal{B}$ given by

$$j_t(f) = f(X_t),$$

for all $f \in \mathcal{A}$, $t \geq 0$.

We now turn into the noncommutative framework. We start by defining a *quantum dynamical semigroup.*

Introduced by physicists during the 1970s, quantum dynamical semigroups (QDS) provide a suitable mathematical framework for studying the evolution of open systems. Typically, an open quantum system involves a dissipative effect modeled via the mutual interaction of different subsystems. One commonly distinguishes between at least the "free system" and the "reservoir."

In general a QDS can be defined over an arbitrary von Neumann algebra, as follows:

Definition 3.1. A *Quantum dynamical semigroup* (QDS) of a von Neumann algebra $\mathcal{A}$ is a weakly*-continuous one-parameter semigroup $(\mathcal{T}_t)_{t\geq 0}$ of completely positive linear normal maps of $\mathcal{A}$ into itself which preserve the identity. In addition, it is assumed that $\mathcal{T}_0$ coincides with the identity map I.

The class of semigroups defined over the von Neumann algebra $\mathcal{A} = \mathcal{B}(h)$ of all bounded operators over a given complex separable Hilbert space h is better known. In particular, several results on the form of the infinitesimal generator of these QDS are available (see, e.g., [31], [13], [27]). We denote by $\mathcal{L}$ the *infinitesimal generator* of the semigroup $\mathcal{T}$, whose domain is given by the set $D(\mathcal{L})$ of all $X \in \mathcal{B}(h)$ for which the w^*-limit of $t^{-1}(\mathcal{T}_t(X) - X)$ exists when $t \to 0$, and we define $\mathcal{L}(X)$ as such a limit.

To understand the form of the generator, we consider a particular case of QDS.

Definition 3.2. A quantum dynamical semigroup $\mathcal{T}$ is called *uniformly continuous* if

$$\lim_{t\to 0} \|\mathcal{T}_t - \mathcal{T}_0\| = 0.$$

From the general theory of semigroups it follows that a QDS is uniformly continuous if and only if its generator $\mathcal{L}$ is a bounded operator. Within this framework the canonical form of a generator has been obtained first by Gorini, Kossakowski and Sudarshan in the finite dimensional case, and extended later by Lindblad to a general Hilbert space in [31], a celebrated result which we recall below as given by Parthasarathy ([39], Theorem 30.16).

Theorem 3.3. [Lindblad] *Let a uniformly continuous quantum dynamical semigroup on the algebra $\mathcal{B}(h)$ of a complex separable Hilbert space h be given. Let ρ be any state in h. Then there exists a bounded selfadjoint operator H and a sequence $(L_k)_{k\in\mathbb{N}}$ of elements in $\mathcal{B}(h)$ that satisfy the following:*

(1) $\mathrm{tr}\rho \mathrm{L_k} = 0$ *for each k;*

(2) $\sum_k L_k^* L_k$ *is a strongly convergent sum;*

(3) If $\sum_k |c_k|^2 < \infty$ *and* $c_0 + \sum_k c_k L_k = 0$ *for scalars* c_k*, then* $c_k = 0$ *for all k;*

(4) The generator $\mathcal{L}$ of the semigroup admits the representation

$$\mathcal{L}(X) = i[H, X] - \frac{1}{2}\sum_k (L_k^* L_k X - 2L_k^* X L_k + X L_k^* L_k),$$

for all $X \in \mathcal{B}(h)$.

This result has been extended by Davies (see [13]) to a class of QDS with unbounded generators.

Generators of QDS commonly appear in physics articles in their *predual* form. That is, given the von Neumann algebra $\mathcal{A} = \mathcal{B}(h)$ its *predual algebra* consists of $\mathcal{A}_* = \mathcal{I}^1(h)$, the algebra of trace class operators. A quantum dynamical semigroup $\mathcal{T}$ induces a *predual* semigroup $\mathcal{T}_*$ on $\mathcal{A}_*$ given by

$$\mathrm{tr}\,(\mathcal{T}_{*t}(Y)X) = \mathrm{tr}\,(Y\mathcal{T}_t(X)),$$

for any $Y \in \mathcal{A}_*$, $X \in \mathcal{A}$.

The generator of the predual semigroup is denoted by $\mathcal{L}_*$. What is usually called a *master equation* in open quantum systems is referred to as the relation between the predual semigroup and its generator, written in the form

$$\frac{d}{dt}\rho_t = \mathcal{L}_*(\rho_t),$$

where $\rho_t = \mathcal{T}_{*t}(\rho)$, for any $t \geq 0$, ρ being a *state*, that is, an element $\rho \in \mathcal{A}_*$ with unitary trace.

It is worth noting that in general a QDS is not a *-homomorphism of algebras. Such a property concerns quantum flows and the concept of *dilation* which we now make precise.

Definition 3.4. A *dilation* of a given QDS is a system $(\mathcal{B}, \mathbb{E}, (\mathcal{B}_t, \mathbb{E}_t, j_t)_{t\geq 0})$ where

1. $\mathcal{B}$ is a von Neumann algebra with a given state $\mathbb{E}$;
2. $(\mathcal{B}_t)_{t\geq 0}$ is an increasing family of von Neumann subalgebras of $\mathcal{B}$;
3. For any $t \geq 0$, $\mathbb{E}_t$ is a conditional expectation from $\mathcal{B}$ onto $\mathcal{B}_t$, such that for all $s, t \geq 0$, $\mathbb{E}_s\mathbb{E}_t = \mathbb{E}_{s\wedge t}$;
4. All maps $j_t : \mathcal{A} \to \mathcal{B}_t$ are *-homomorphisms which preserve identity and satisfy the Markov property
$$\mathbb{E}_s \circ j_t = j_t \circ \mathcal{T}_{t-s}.$$

   $J = (j_t)_{t\geq 0}$ is known as a *quantum Markov flow* associated to a given QDS.

We call the structure $\mathfrak{B} = (\mathcal{B}, \mathbb{E}, (\mathcal{B}_t, \mathbb{E}_t)_{t\geq 0})$ a *quantum stochastic basis.*

The canonical form of a quantum Markov flow is given by $j_t(X) = V(t)^*XV(t)$, $(t \geq 0)$, where $V(t) : \mathcal{A} \to \mathcal{B}_t$ is a *cocycle* with respect to a given family of *time-shift* operators $(\theta_t)_{t\geq 0}$. To be more precise, we have the following:

Definition 3.5. Given a quantum stochastic basis $\mathfrak{B}$, a family $(\theta_t)_{t\geq 0}$ of *-homomorphisms of $\mathcal{B}$ is called a *covariant shift* if

- $\theta_0(Y) = Y$,
- $\theta_t(\theta_s(Y)) = \theta_{t+s}(Y)$,
- $\theta_t^*(\theta_t(Y)) = Y$,
- $\theta_t(\mathbb{E}_0(\theta_s(Y))) = \mathbb{E}_t(\theta_{t+s}(Y))$,

for any $Y \in \mathcal{B}$. A family $(V(t))_{t\geq 0}$ of elements in $\mathcal{B}$ is a *left cocycle* (resp. *right cocycle*) with respect to a given covariant shift whenever

$$V(t+s) = V(s)\theta_s(V(t)), \text{ (resp. } V(t+s) = \theta_s(V(t))V(s)), \ (s, t \geq 0). \quad (3.1)$$

Example 1. According to the above definition, the quantum dynamical semigroup associated to the Schrödinger equation is

$$\mathcal{T}_t(X) = U^*(t)XU(t),$$

where $U(t) = e^{-itH}$ and H is a selfadjoint operator in $\mathcal{B}(h)$. Here $\mathcal{T}_t$ is itself a *-homomorphism; U is a fortiori a cocycle with respect to the trivial shift $\theta_t = \theta_0 = I$. The differential equation satisfied by U is simply $dU(t) = -iHU(t)dt$ and the generator of $\mathcal{T}$ is given by

$$\mathcal{L}(X) = i[H, X], \ (X \in \mathcal{B}(h)).$$

Example 2. Let there be given a filtered probability space $(\Omega, \mathcal{F}, (\mathcal{F}_t)_{t\geq 0}, \mathbb{P})$ where a classical Brownian motion $W = (W_t)_{t\geq 0}$ is defined and a complex separable Hilbert space h. For instance, let Ω be the space $C(\mathbb{R}_+, \mathbb{R})$ of continuous real functions endowed with the Wiener measure $\mathbb{P}$, $W_t(\omega) = \omega(t)$, for any $\omega \in \Omega, \ t \geq 0$. Consider the stochastic differential equation

$$d\psi_t = (-iH + K)\psi_t dt + L\psi_t dW_t; \ \psi_0 = \psi, \tag{3.2}$$

where $H = H^*, K = K^*, L$ are elements of $\mathcal{B}(h)$ and ψ is a fixed unitary vector of h.

This equation has a unique solution $(\psi_t)_{t\geq 0}$, with $\psi_t \in L^2_h(\Omega, \mathcal{F}, \mathbb{P})$, $(t \geq 0)$. To have $\mathbb{E}(\|\psi_t\|^2) = 1$ for all $t \geq 0$, K and L need to satisfy additional conditions. Indeed,

$$\begin{aligned} d\langle\psi_t, \psi_t\rangle &= \langle d\psi_t, \psi_t\rangle + \langle\psi_t, d\psi_t\rangle + \langle d\psi_t, d\psi_t\rangle \\ &= \langle\psi_t, (2K + L^*L)\psi_t\rangle dt + \langle\psi_t, (L^* + L)\psi_t\rangle dW_t. \end{aligned}$$

Thus, $d\mathbb{E}(\langle\psi_t, \psi_t\rangle) = 0$ if and only if

$$K = -\frac{1}{2}L^*L. \tag{3.3}$$

Furthermore, let $Z(t) = (-iH - \frac{1}{2}L^*L)t + LW_t$. This is an operator-valued semimartingale and the stochastic Schrödinger equation may be written

$$d\psi_t = dZ(t)\psi_t, \ \psi_0 = \psi.$$

Note that $L^2_h(\Omega, \mathcal{F}, \mathbb{P})$ is isomorphic to $h \otimes L^2_{\mathbb{C}}(\Omega, \mathcal{F}, \mathbb{P})$. And we can define an operator-valued (classical) stochastic process $V : \Omega \times \mathbb{R}_+ \to \mathcal{B}(h)$, such that $V(\cdot, t) : h \mapsto h \otimes L^2_{\mathbb{C}}(\Omega, \mathcal{F}_t, \mathbb{P})$, for any $t \geq 0$, which associates to each unitary $\psi \in h$ the unique solution ψ_t of (3.2). This operator-valued process may be interpreted as a family $(V(t))_{t\geq 0}$ of applications belonging to the algebra $L^\infty_{\mathcal{B}(h)}(\Omega, \mathcal{F}, \mathbb{P})$, that is $V : \mathbb{R}_+ \to L^\infty_{\mathcal{B}(h)}(\Omega, \mathcal{F}, \mathbb{P})$, given by $t \mapsto V(\cdot, t)$. The process $V(t)$ is given as the solution to the (right) operator-valued linear stochastic differential equation

$$dV(t) = (-iH - \frac{1}{2}L^*L)V(t)dt + LV(t)dW_t = dZ(t)V(t), \ V(0) = I. \tag{3.4}$$

We now consider the classical shift operator $\theta_t : \Omega \to \Omega$, given by $\theta_t(\omega)(s) = \omega(t+s)$ for any $s, t \geq 0$, $\omega \in \Omega$. This induces a shift, denoted by θ_s, on the elements $Y \in \mathcal{B} = L^{\infty}_{\mathcal{B}(h)}(\Omega, \mathcal{F}, \mathbb{P})$ given by

$$\theta_s(Y) = Y \circ \theta_s.$$

This shift is covariant with respect to the family $(\mathbb{E}(\cdot/\mathcal{F}_t))_{t\geq 0}$, i.e.,

- $\theta_0(Y) = Y$, for any $Y \in \mathcal{B}$;
- $\theta_t(\theta_s(Y)) = \theta_{t+s}(Y)$;
- $\theta_t^*(\theta_t(Y)) = Y$;
- $\theta_t(\mathbb{E}(\theta_s(Y))) = \mathbb{E}(\theta_{t+s}(Y)/\mathcal{F}_t)$.

$V(t)$ is then a right cocycle with respect to this family of applications since, as in the classical case, the following relation is satisfied:

$$V(\omega, t+s) = V(\theta_s(\omega), t)V(\omega, s).$$

We now define the corresponding flow as

$$j_t(X) = V(t)^* X V(t), \quad (X \in \mathcal{B}(h)).$$

And we obtain

$$\begin{aligned} dj_t(X) &= dV(t)^* X V(t) + V(t)^* X dV(t) + dV(t)^* X dV(t) \\ &= V(t)^* dZ(t)^* X V(t) + V(t)^* X dZ(t) V(t) \\ &\quad + V(t)^* dZ(t)^* X dZ(t) V(t) \\ &= j_t(\mathcal{L}(X))dt + j_t(\alpha(X))dW_t, \end{aligned} \tag{3.5}$$

where

$$\mathcal{L}(X) = i[H, X] - \frac{1}{2}(L^*LX - 2L^*XL + XL^*L); \tag{3.6}$$

$$\alpha(X) = XL + L^*X. \tag{3.7}$$

Now the associated quantum dynamical semigroup is given by

$$\mathcal{T}_t(X) = \mathbb{E}(j_t(X)),$$

which has a generator obtained from the structure equation (3.5), and given by (3.6).

Thus, stochastic Schrödinger equations are naturally included within the framework of quantum dynamical semigroups. They represent a *classical* dilation of the QDS whose generator is (3.6). The adjective "classical" is used here for the probabilistic model used to dilate the semigroup. A given QDS may have many different dilations based on classical or quantum noise.

This example may be generalized as follows: take a collection $(W^j)_{j\geq 1}$ of independent Brownian motions, and operators $L_j \in \mathcal{B}(h)$ such that $\sum_j L_j^* L_j$ converges in $\mathcal{B}(h)$. Consider the stochastic Schrödinger equation:

$$d\psi_t = (-iH - \frac{1}{2}\sum_j L_j^* L_j)\psi_t dt + \sum_j L_j dW_t^j \psi_t, \; \psi_0 = \psi. \tag{3.8}$$

The corresponding associated QDS has a generator of the form

$$\mathcal{L}(X) = i[H, X] - \frac{1}{2}\sum_j (L_j^* L_j X - 2L_j^* X L_j + X L_j^* L_j). \tag{3.9}$$

We profit from this example to explain a useful notation for stochastic differentials due to Belavkin. We introduce noise as follows:

$$\Lambda_0^0(t) = t, \tag{3.10}$$

$$\Lambda_j^0(t) = W_t^j = \Lambda_0^j(t), \tag{3.11}$$

for all $j \geq 1$, $t \geq 0$. These are real scalar noise, thus the expression $\Lambda_m^{\ell *}$ has a trivial meaning. However, due to the further extension to operator noise, we adopt the convention $\Lambda_m^{\ell *} = \Lambda_\ell^m$. We use the customary convention of probabilists for writing characteristic functions: $1_{\{\dots\}}$ means that we write 1 if condition $\{\dots\}$ is satisfied and 0 otherwise.

With this notation Itô's formula is written in a compact differential form as

$$d\Lambda_m^k d\Lambda_j^\ell = d\Lambda_j^\ell d\Lambda_m^k = 1_{\{j=k>0,\ell=m=0\}} d\Lambda_0^0. \tag{3.12}$$

Moreover, if we denote $L_j^0 = L_j = L_0^j$ for all $j \geq 1$ and $L_0^0 = -iH - \frac{1}{2}\sum_j L_j^* L_j$, $L_j^k = 0$ for $j, k \geq 1$ the equation of the cocycle becomes

$$dV(t) = dZ(t)V(t),$$

where

$$dZ(t) = \sum_{j,k=0}^{\infty} L_j^k d\Lambda_k^j(t). \tag{3.13}$$

Furthermore, to obtain the structure equation for the quantum flow $j_t(X) =$

$V(t)^*XV(t)$ we use the above equation, and (3.12) yields

$$\begin{aligned}
dj_t(X) &= d(V(t)^*XV(t)) \\
&= V(t)^*dZ(t)^*XV(t) \\
&+ V(t)^*XdZ(t)V(t) \\
&+ V(t)^*dZ(t)^*XdZ(t)V(t) \\
&= j_t(\sum_{\ell,m} L_m^{\ell\,*} d\Lambda_m^{\ell}(t)X) \\
&+ j_t(\sum_{\ell,m} XL_m^{\ell} d\Lambda_m^{\ell}(t)) \\
&+ j_t(\sum_{\ell,m,j,k} L_k^{j\,*} d\Lambda_k^{j}(t)XL_m^{\ell} d\Lambda_m^{\ell}(t)).
\end{aligned}$$

As a result,

$$dj_t(X) = \sum_{\ell,m=0}^{\infty} j_t(\theta_\ell^m(X))d\Lambda_m^{\ell}(t), \tag{3.14}$$

for all $X \in \mathcal{B}(h)$, where $(\theta_\ell^m)_{\ell,m}$ is called the family of *structure maps* of the flow, given by

$$\theta_k^{\ell}(X) = L_m^{\ell\,*}X + XL_\ell^m + \sum_{k=1}^{\infty} L_k^{\ell\,*}XL_k^m, \tag{3.15}$$

for any $\ell, m \in \mathbb{N}$. Since the QDS is given by $\mathcal{T}_t(X) = \mathbb{E}(j_t(X))$, it is worth noting that we recover the generator of the semigroup via the structure map $\theta_0^0(X)$ which is associated to the noise $d\Lambda_0^0(t) = dt$ because other noise is projected by $\mathbb{E}$ on 0:

$$\begin{aligned}
\mathcal{L}(X) = \theta_0^0(X) &= L_0^{0\,*}X + XL_0^0 + \sum_{k=1}^{\infty} L_k^{0\,*}XL_k^0 \\
&= i[H,X] - \frac{1}{2}\sum_j (L_j^*L_jX - 2L_j^*XL_j + XL_j^*L_j).
\end{aligned}$$

Example 3. Consider again a classical probability space $(\Omega, \mathcal{F}, \mathbb{P})$ where we suppose defined two independent Poisson processes N^a, N^d with respective intensities λ_a, λ_d. In addition, we take a unitary operator U defined on a given complex separable Hilbert space h. Consider the stochastic differential equation

$$dV(t) = \{(U - I)dN_t^a + (U^* - I)dN_t^d\}V(t), \ V(0) = I, \tag{3.16}$$

This equation provides a unitary cocycle V. Moreover, using Belavkin notation we put

$$L_1^1 = U - I, \ L_2^2 = (U^* - I); \ L_k^j = 0, \text{ otherwise},$$

$$\Lambda_1^1 = N^a, \ \Lambda_2^2 = N^d.$$

The Itô multiplication rule becomes

$$d\Lambda_m^k d\Lambda_j^\ell = d\Lambda_j^\ell d\Lambda_m^k = 1_{\{j=k=\ell=m>0\}} d\Lambda_\ell^\ell \tag{3.17}$$

With this choice of notation, the computation of the structure maps gives

$$\theta_1^1(X) = U^*XU - X; \theta_2^2(X) = UXU^* - X, \ (X \in \mathcal{B}(h))$$

that is, the equation for the flow is

$$dj_t(X) = j_t(U^*XU - X)dN_t^a + j_t(UXU^* - X)dN_t^d.$$

Since $\mathbb{E}(N_t^a) = \lambda_a t, \ \mathbb{E}(N_t^d) = \lambda_d t$, the generator of the QDS has the form

$$\mathcal{L}(X) = \lambda_a(U^*XU - X) + \lambda_d(UXU^* - X), \ (X \in \mathcal{B}(h)). \tag{3.18}$$

In particular, take $h = l^2(\mathbb{Z})$ with its canonical orthonormal basis (e_n). We define the right shift as $Se_n = e_{n+1}$, and the number operator $Ne_n = ne_n$, $(n \in \mathbb{Z})$. Take $U = S$ and $X = f(N)$, where $f(N)e_n = f(n)e_n$ for all $n \in \mathbb{Z}$, f being any bounded function defined on the integers. Then (3.18) gives

$$\mathcal{L}(f(N)) = \lambda_a(f(N+1) - f(N)) + \lambda_d(f(N-1) - f(N)),$$

which coincides with the generator of a birth and death process.

Example 4. Take E to be a compact space and call $\mathcal{E}$ its Borel σ-field and a probability space $(\Omega, \mathcal{F}, \mathbb{P})$. Given a finite measure μ on $(E, \mathcal{E})$ we consider a marked Poisson process with intensity μ which is characterized by a double sequence of random variables $(\xi_n, T_n)_{n\geq 1}$ taking values in $E \times \mathbb{R}_+$ such that

$$N(A, [0, t]) = \sum_{n\geq 1} \delta_{(\xi_n, T_n)}(A \times]0, t]),$$

has a Poisson distribution with intensity parameter $\mu(A)$, for all $A \in \mathcal{E}, t \geq 0$.

Consider a family $(U(x))_{x\in E}$ of unitary operators on a given complex separable Hilbert space h and a selfadjoint operator H. Moreover, we assume that the map $x \mapsto U(x)$ from E into $\mathcal{B}(h)$ is measurable (strong or weak measurability are equivalent in this case). We consider a Schrödinger equation where the dynamics is perturbed by random kicks as follows:

$$d\psi_t = -iH\psi_t dt + \sum_{n\geq 1}(U(\xi_n) - I)\psi_{T_n} 1_{\{T_n \leq t < T_{n+1}\}}. \tag{3.19}$$

The above equation may be written for a cocycle as

$$dV(t) = \left[-iHdt + \int_E (U(x) - I)N(dx, dt)\right] V(t). \tag{3.20}$$

To justify these expressions, we verify the integrability of the map $x \mapsto U(x) - I$ with respect to the Poisson process. Indeed, for any $t \geq 0$,

$$\mathbb{E}\left(\int_E \|U(x) - I\| N(dx,]0,t])\right) \leq 2\mu(E)\, t < \infty.$$

Proceeding as we did in the previous examples, we examine the differential equation satisfied by the flow $j_t(X) = V(t)^* X V(t)$, $(t \geq 0,\ X \in \mathcal{B}(h))$. Here we have $dZ(t) = -iHdt + \int_E (U(x) - I)N(dx, dt)$, therefore

$$dj_t(X) = j_t(dZ(t)^* X + X dZ(t) + dZ(t)^* X dZ(t)).$$

We use Itô's rule, which now becomes $N(\{x\}, dt)N(\{y\}, dt) = \delta_{x,y} N(\{y\}, dt)$, and it follows that

$$dj_t(X) = j_t(i[H, X])dt + \int_E j_t(U(x)^* X U(x) - X)N(dx, dt).$$

Finally, taking expectations

$$\mathcal{T}_t(X) = \mathbb{E}\left(\int_0^t j_s(i[H, X] + \int_E (U(x)^* X U(x) - X)\mu(dx))ds\right)$$

means that the generator $\mathcal{L}$ of the QDS has the form

$$\mathcal{L}(X) = i[H, X] + \int_E (U(x)^* X U(x) - X)\mu(dx), \ (X \in \mathcal{B}(h)). \tag{3.21}$$

Note that the master equation associated to (3.21) provides an example of a quantum Boltzmann equation. Indeed, the master equation is simply

$$\frac{\partial \rho_t}{\partial t} = \mathcal{L}_*(\rho_t), \ \rho_0 = \rho \in \mathcal{I}^1(h).$$

Now, introduce a *collision map*

$$\kappa_*(\rho) = \int_E (U(x)\rho U(x)^* - \rho)\mu(dx).$$

This allows one to write the predual generator as $\mathcal{L}_*(\rho) = -i[H, \rho] + \kappa_*(\rho)$, and the master equation becomes

$$\frac{\partial \rho_t}{\partial t} + i[H, \rho_t] = \kappa_*(\rho_t), \ \rho_0 = \rho. \tag{3.22}$$

The above examples provide a partial view of quantum dynamical semigroups; in all these cases the generators are bounded, which is not satisfactory from the point of view of physical applications. Moreover, the dilations have been built in a classical method via Wiener or Poisson processes. More general dilations, i.e., those obtained with quantum noise are more suitable for the description of open quantum systems. Quantum noise appears naturally within the framework of Fock spaces. Numerous authors (see for instance [34]) have stressed the main advantage of a (boson) Fock space: that structure supports both canonical commutation relations (CCR) and a theory of stochastic integration with respect to quantum noise, providing a noncommutative version of the Itô's algebra for differentials.

The study of linear quantum stochastic differential equations with unbounded coefficients have been done by several authors. Fagnola in [14] established a useful criterion for the existence and uniqueness of solutions to equations of the form

$$dV(t) = V(t) \sum_{\ell,m} L_{\ell}^{m} d\Lambda_{m}^{\ell}(t),$$

where the processes Λ_m^ℓ are quantum noise. We will use this result in the next section. Notice that $V(t)$ appears here on the left of the right-hand side of the equation. This is done to simplify the handling of domains which appear as a major problem related to unbounded coefficients. As a result, one uses this to write flows like $j_t(X) = V(t)XV(t)^*$ instead of $V(t)^*XV(t)$.

We will show a particular example of a noncommutative dilation inspired from quantum optics. This dilation uses quantum noise and rephrases most of a joint work with Fagnola and Saavedra on the subject. The interested reader is referred to [16], [17], [21] for further details.

4 The open system of quantum optics

The statement regarding master equations of quantum optics follows some general principles which we make explicit:

- The device is based upon the emission of photons due to the interaction of an electromagnetic field and a collection of atoms. The electromagnetic field acts within a *cavity*,
- The atoms are injected into the cavity at different times. They interact with the electromagnetic field but they do not interact with each other,
- The energy level of each atom is affected by the interaction with the field.

This is the the basic framework for a microscopic approach to laser and maser descriptions (see [33]).

4.1 A classical radiation model due to Einstein

We follow Einstein's ideas about the atom–radiation interaction (see for instance [37]), expressed in terms of classical stochastic processes.

To begin, we assume that n identical atoms are confined in a closed cavity, with two relevant bounded-state energy levels $E_k = k\hbar\varpi$ ($k = 0, 1$), where ϖ denotes frequency. There is an external source of electromagnetic energy. The average energy density $E(\mathfrak{w})$ is the sum of the energies due to thermal radiation and external sources. Three effects takes place. First, an atom whose energy decays from the energy level 1 to 0 emits a photon of energy $\hbar\mathfrak{w}$. Second, for an atom at level E_0, absorption of energy eventually leads to level E_1; the rate of this type of transition is proportional to the electromagnetic energy $E(\mathfrak{w})$ present in the cavity. Third, Einstein proposed including the so-called *stimulated emission*, which produces a transition from E_1 to E_0 at a rate proportional to $E(\mathfrak{w})$.

We now let $n \to \infty$, the number of atoms at a given energy level is given by the birth and death process X. Assume, for instance, that X_t counts the number of atoms at level E_0 during the interval $[0, t]$. Call α_n (resp. β_n) the parameter of the exponential law which governs the time for a transition to level E_0 (resp. to level E_1) when the process is at state n. Consider the Markovian transition matrix

$$Q(n, m) = \begin{cases} \frac{\alpha_n}{\alpha_n+\beta_n}, & \text{if } m = n + 1 \\ \frac{\beta_n}{\alpha_n+\beta_n}, & \text{if } m = n - 1 \\ 0, & \text{otherwise.} \end{cases}$$

Then there exists a probability space $(\Omega, \mathcal{F}, \mathbb{P})$ and a Markov jump process X with states in $\mathbb{N}$ that is characterized by a random sequence $(\xi_n, T_n)_n$ of elements in $\mathbb{N} \times \mathbb{R}_+$ such that

- $(\xi_n)_n$ is a Markov chain with transition Q with respect to the family of σ-fields $\mathcal{F}_n = \sigma(\xi_k, T_k;\ k \leq n)$.

- Conditionally on $\mathcal{F}_n$, $T_{n+1} - T_n$ is independent of ξ_{n+1} for all $n \in \mathbb{N}$, distributed according to the exponential law of parameter $\alpha_n + \beta_n$.

And X_t is given by

$$X_t = \sum_{n\in\mathbb{N}} \xi_n 1_{\{T_n \leq t \leq T_{n+1}\}}.$$

From the previous discussion of Einstein's argument, the parameters have the form $\alpha_n = \lambda_n + R_n E(\mathfrak{w})$ and $\beta_n = \mu_n E(\mathfrak{w})$. Thus, the generator of the Markov process X can be written in the form

$$Lf(n) = (\lambda_n + R_n E(\mathfrak{w}))(f(n+1) - f(n)) - \mu_n E(\mathfrak{w})(f(n) - f(n-1)), \tag{4.1}$$

for any bounded function f defined on $\mathbb{N}$ and all $n \in \mathbb{N}$.

This is too simplified a model, since it does not contain quantum noise due to spontaneous emission; this has motivated serious criticism among physicists (see for instance the analysis done in [37]) even though Einstein's probabilistic concepts support our intuition when one is concerned with more sophisticated models.

4.2 A step forward for a noncommutative model

We are now interested in deriving an equation describing the evolution of the field state. What we mean by the *state at time t* is a positive trace class operator ρ_t with unit trace defined over an initial Hilbert space h. This type of equation is named after Langevin. The first known research on such equations in quantum theory is due to Senitzky, and further used by Haken and Lax to study quantum optics. The Master Equation obtained by Scully and Lamb by *coarse–grained approximation* [46], [45] inspired my joint research with Fagnola and Saavedra (see [16], [21]). We recall below this result.

Let me presemt some intuitive ideas before we proceed with the equations. Following the ideas of Einstein expressed in a noncommutative manner, we need operators that could represent the rising and lowering of energy levels. This can be done via creation and annihilation operators for which we need a suitable choice of the initial space is $h = l^2(\mathbb{Z})$ with its canonical basis (e_n).

Definition 4.1. We define the following fundamental operators on h:

1. *The annihilation operator*

$$ae_n = \begin{cases} \sqrt{n}\, e_{n-1} & , \text{ if } n > 0 \\ 0 & , \text{ if } n \leq 0, \end{cases} \tag{4.2}$$

for all $n \in \mathbb{N}$. The domain of a is $D(a) = \{x = (x_n)_n \in h : \sum_n |n||x_n|^2 < \infty\}$

2. *The creation operator*

$$a^\dagger e_n = \begin{cases} \sqrt{n+1}\, e_{n+1} & , \text{ if } n \geq 0, \\ 0 & , \text{ if } n < 0, \end{cases} \tag{4.3}$$

for all $n \in \mathbb{N}$. The domain of $a^\dagger$ is also $D(a^\dagger) = D(a)$.

3. *The number operator*

$$Ne_n = \begin{cases} ne_n & , \text{ if } n > 0, \\ 0 & , \text{ if } n \leq 0, \end{cases} \tag{4.4}$$

for all $n \in \mathbb{N}$. The domain of N is $D(N) = \{x = (x_n)_n \in h : \sum_n |n|^2|x_n|^2 < \infty\}$.

4. *The right-shift operator*

$$Se_n = e_{n+1}, \ (n \in \mathbb{N}). \tag{4.5}$$

These operators are related as follows:

$$a^\dagger = N^{1/2}S = S(N+1)^{1/2}, \tag{4.6}$$

$$a = (N+1)^{1/2}S^* = S^* N^{1/2}. \tag{4.7}$$

Furthermore, given any function $\varphi : \mathbb{N} \to \mathbb{C}$, we define a function of the number operator as follows:

$$\varphi(N)\, e_n = \begin{cases} \varphi(n)e_n & , \text{ if } n > 0 \\ \varphi(0)e_n & , \text{ if } n \leq 0 \end{cases} \tag{4.8}$$

The master equation below, derived in [16], was obtained according to the phenomenological approach described at the beginning of this section. We omit its derivation and refer the reader to [16].

$$\begin{aligned} \frac{d}{dt}\rho_t &= \frac{\mu^2}{2}(-a^\dagger a\rho_t + 2a\rho_t a^\dagger - \rho_t a^\dagger a) \\ &+ \frac{\lambda^2}{2}(-aa^\dagger\rho_t + 2a^\dagger\rho_t a - \rho_t aa^\dagger) \\ &+ R^2(\varphi_1(N)\rho_t\varphi_1(N) - \rho(t)) \\ &+ R^2 S\varphi_2(N)\rho_t\varphi_2(N)S^*, \end{aligned} \tag{4.9}$$

where $\lambda = \sqrt{\gamma_c n_T}$, $\mu = \sqrt{\gamma_c(n_T+1)}$, (with constants $\gamma_c, n_T > 0$), where φ_1, φ_2 are given by

$$\varphi_1(N) = \cos(\phi\sqrt{N+1}), \tag{4.10}$$
$$\varphi_2(N) = \sin(\phi\sqrt{N+1}). \tag{4.11}$$

Equivalently, we have a semigroup of operators $(\mathcal{T}_t)$ which has a generator $\mathcal{L}(X)$ defined over a domain $D(\mathcal{L})$ to be made precise later, decomposed in two pieces:

$$\mathcal{L}(X) = \mathcal{L}_0(X) + \mathcal{L}_{ga}(X), \ (X \in D(\mathcal{L}), \tag{4.12}$$

where $\mathcal{L}_0$ is associated to the free dynamics, that is without considering the action of an external electromagnetic field, and $\mathcal{L}_{ga}$ describes the gain induced by the interaction with the external field.

$$\begin{aligned} \mathcal{L}_0(X) &= \frac{\mu^2}{2}(-NX + 2N^{1/2}SXS^*N^{1/2} - XN) \\ &+ \frac{\lambda^2}{2}(-(N+1)X + 2(N+1)^{1/2}S^*XS(N+1)^{1/2} - X(N+1)) \end{aligned}$$

$$\mathcal{L}_{ga}(X) = R^2\left(\varphi_1(N)X\varphi_1(N) + \varphi_2(N)S^*XS\varphi_2(N) - X\right).$$

If we follow the intuitive ideas of Einstein, the generator $\mathcal{L}$ should be associated to a birth and death process. We will explore this idea later. Birth and death generators are *included as a particular case of this general model* when restricted to a suitable abelian algebra.

4.3 The associated quantum stochastic differential equation

We now come to dilate the QDS characterized by the generator (4.12) to construct an associated quantum Markov flow. This will be done by introducing appropriate quantum noise and stochastic differential equations which will provide us with a Markov cocycle leading to the flow.

We let $\mathcal{H}$ be the Hilbert space defined as the tensor product

$$\mathcal{H} = h \otimes \Gamma(L^2(\mathbf{R}_+, \mathbb{C}^4)),$$

where $h = l^2(\mathbb{Z})$ is the initial space mentioned earlier and $\Gamma(L^2(\mathbf{R}_+, \mathbb{C}^4))$ is the usual Boson–Fock space of $\mathbb{C}^4$-valued functions.

We work now with operators defined on $\mathcal{H}$. This leads to considering $\mathbb{C}^4$-valued functions. To make explicit the action over each component we take the canonical basis $(u_k)_{k=1}^4$ of $\mathbb{C}^4$. As usual, $|u_l\rangle\langle u_m|$ denotes the projection $|u_l\rangle\langle u_m|z = \langle u_m, z\rangle u_l$ for any $z \in \mathbb{C}$. Moreover, given any $t > 0$, the multiplication operator by the characteristic function $1_{]0,t[}$, denoted by the same symbol, defines a projection on the space $L^2(\mathbf{R}_+, \mathbb{C})$. As a result, since $L^2(\mathbf{R}_+, \mathbb{C}^4)$ is isomorphic to $L^2(\mathbf{R}_+, \mathbb{C}) \otimes \mathbb{C}^4$, the operator $1_{]0,t[} \otimes |u_l\rangle\langle u_m|$ is a projection on $L^2(\mathbf{R}_+, \mathbb{C}^4)$. The usual notation $e(f) \in \Gamma(L^2(\mathbf{R}_+, \mathbb{C}^4))$ will be adopted for the exponential vector corresponding to the test function $f \in L^2(\mathbf{R}_+, \mathbb{C}^4)$, that is

$$e(f) = \sum_{n\geq 0} \frac{1}{\sqrt{n!}} f^{\odot n}.$$

In the next section we will write elements of the form $u \otimes e(f)$, $(u \in h,\ f \in L^2(\mathbf{R}_+, \mathbb{C}^4))$ by $ue(f)$. On the other hand, given an operator X on h, we extend this to $\mathcal{H}$ as $X \otimes I$, where I is the identity operator on the Fock space. In addition, we introduce the canonical projection $E : \mathcal{H} \to h$ defined as $Eue(f) = ue(0)$ (where h is identified with the space $\{ue(0) :\ u \in h\}$).The above projection of $\mathcal{H}$ induces a projection $\mathbb{E}$ on the algebra as follows:

$$\mathbb{E}(Z) = EZE \in \mathcal{B}(h), \tag{4.13}$$

for any $Z \in \mathcal{B}(\mathcal{H})$.

For all $t \geq 0$ consider the operators on $\Gamma(L^2(\mathbb{R}_+; \mathbb{C}^4))$ defined by

$$\begin{aligned} \Lambda_m^l(t)e(f) &= \Lambda(1_{]0,t[} \otimes |u_l\rangle\langle u_m|)e(f), \\ &= -i\frac{d}{d\varepsilon} e(\exp(i\varepsilon 1_{]0,t[} \otimes |u_l\rangle\langle u_m|) f)\Big|_{\varepsilon=0}. \end{aligned} \tag{4.14}$$

Λ is known as the *number operator.*

$$\begin{aligned} \Lambda_m^0(t)e(f) &= A^\dagger(1_{]0,t[} \otimes \langle u_m|)e(f) \\ &= \frac{d}{d\varepsilon} e(f + \varepsilon 1_{]0,t[} u_m)\Big|_{\varepsilon=0}, \end{aligned} \tag{4.15}$$

$A^\dagger$ is called the *creation operator*.

$$\begin{aligned}\Lambda_0^l(t)e(f) &= A(1_{]0,t[} \otimes |u_l\rangle) \\ &= \langle u_l 1_{]0,t[}, f\rangle e(f).\end{aligned} \tag{4.16}$$

A is called the *annihilation operator*.

$$\Lambda_0^0(t)e(f) = te(f), \tag{4.17}$$

which is simply tI.

For these operators Itô's table holds:

$$d\Lambda_m^k(t)d\Lambda_j^\ell(t) = 1_{\{j=k>0\}}d\Lambda_m^\ell(t). \tag{4.18}$$

4.4 Master equation for the quantum cocycle

Definition 4.2. A *quantum cocycle* $V = (V(t))_{t\geq 0}$ connected to the semigroup $\mathcal{T}$ is a family of contractions on $\mathcal{H}$ such that the semigroup is represented in the form

$$\mathcal{T}_t(X) = \mathbb{E}(V(t)XV(t)^*) = EV(t)(X \otimes I)V(t)^*E,$$

for all $t \geq 0$ and $X \in \mathcal{B}(h)$.

We first derive formally the equation for the cocycle V from the master equation of quantum optics. After that we will prove the existence and uniqueness of solutions to the stated equation.

The coefficients of our equation are taken as the closure of the following operators defined over the domain D which is the linear span of the basis $(e_m)_m$ of h.

$$L_0^1 = -\mu N^{1/2},\ L_0^2 = -\lambda(N+1)^{1/2},\ L_0^3 = -R\varphi_1(N),\ L_0^4 = -R\varphi_2(N),$$

$$L_1^0 = \mu N^{1/2}S,\ L_2^0 = \lambda(N+1)^{1/2}S^*,\ L_3^0 = R\varphi_1(N),\ L_4^0 = R\varphi_2(N)S^*,$$

$$L_1^1 = S - I,\ L_2^2 = S^* - I,\ L_4^4 = S^* - I,$$

$$L_0^0 = -\frac{1}{2}(\mu^2 N + \lambda^2(N+1) + R^2),\ L_m^1 = 0 \text{ otherwise.}$$

With the above notation the master equation becomes

$$dV(t) = \sum_{l,m=0}^{4} V(t)L_m^l d\Lambda_l^m(t), \tag{4.19}$$

$$V(0) = I. \tag{4.20}$$

This is our fundamental quantum stochastic differential equation.

Note that the following two conditions are necessary for V to be unitary:

$$(L_m^l + (L_l^m)^* + \sum_{k=1}^{4} L_k^l (L_k^m)^*)u = 0 \ (u \in D), \tag{4.21}$$

$$(L^l_m + (L^m_l)^* + \sum_{k=1}^{4}(L^k_l)^* L^k_m)u = 0, \ (u \in D). \tag{4.22}$$

Since D is a common invariant domain for the operators $(L^l_m)^4_{l,m=0}$ we can apply the main theorem in [14] to prove the existence of a solution to (4.20).

Theorem 4.3. [see [16]] *There exists a unique unitary cocycle V which is a solution to the quantum stochastic differential equation* (4.20) *associated to the master equation in quantum optics. Furthermore, V is strongly continuous on vectors of the form $ue(f)$ with $u \in D$, $f \in L^2(\mathbf{R}_+; \mathbb{C}^4)$.*

The solution to this master equation gives the evolution of all observables of the quantum system. Note that some of them are related to well-known classical stochastic processes. We will particularly concentrate on some of those classical processes later, but we first derive the differential equation satisfied by the flow.

4.5 The quantum Markov flow

The quantum Markov flow in our model is given by the expression

$$j_t(X) = V(t)XV(t)^*, \tag{4.23}$$

for all $t \geq 0$ and $X \in \mathcal{B}(h)$.

This dilation of the quantum dynamical semigroup has the important property of sending any element on the initial algebra into a bigger algebra depending on time. Indeed, fixing $t \geq 0$ and $X \in \mathcal{B}(h)$, the image $j_t(X)$ is an element of the algebra $\mathcal{B}(\mathcal{H}_t)$, where $\mathcal{H}_t$ is the space obtained as

$$\mathcal{H}_t = h \otimes \Gamma(L^2([0,t]; \mathbb{C}^4)).$$

Here again the role played by the Fock space is crucial. It allows extending the notion of *filtration* in classical probability since the total space $\Gamma(L^2([0,\infty[; \mathbb{C}^4))$ may be (isomorphically) expressed as a tensor product of two pieces: that is, $\Gamma(L^2([0,t]; \mathbb{C}^4))$ and $\Gamma(L^2(]t,\infty[; \mathbb{C}^4))$ for any $t \geq 0$. Moreover, the quantum flow is connected to the quantum dynamical semigroup $\mathcal{T}$ via the relation

$$\mathcal{T}_t(X) = \mathbb{E}(j_t(X)),$$

for all $X \in \mathcal{B}(h)$ and $t \geq 0$.

We finish this discussion by stating a corollary of Theorem 4.3.

Corollary 4.4. [see [16]] *For all $X \in \mathcal{B}(h)$ for which $X(D(N)) \subset D(N)$ the quantum Markov process $(j_t(X))_{t\geq 0}$ satisfies the stochastic equation*

$$dj_t(X) = \sum_{\ell,m=0}^{4} j_t(\theta^\ell_m(X))d\Lambda^\ell_m(t), \ j_0(X) = X, \tag{4.24}$$

where the structure maps θ^ℓ_m are given by

$$\theta^\ell_m(X) = L^\ell_m X + X(L^m_\ell)^* + \sum_{k=1}^{4} L^\ell_k X(L^m_k)^*, \quad (0 \le \ell, m \le 4). \tag{4.25}$$

Proof. It suffices to apply Ito's formula to compute the differential of $j_t(X)$. □

Remark. We constructed a fully *quantum dilation* of a given quantum dynamical semigroup whose Lindblad generator $\mathcal{L} = \theta^0_0$ is characterized via the operators L^ℓ_m introduced previously. A natural question arises as to whether we can obtain a dilation with classical noise as in the examples of Section 3?

To analyze this problem, we use classical independent Wiener processes as in Example 2 of Section 3. To distinguish classical noise and processes from their quantum counterpart, we use boldface symbols. Thus, we have here $\mathbf{\Lambda}^0_0 = t$, $\mathbf{\Lambda}^\ell_0 = W^\ell = \mathbf{\Lambda}^0_\ell$, $(1 \le \ell \le 4)$. We next introduce the stochastic differential equation for the cocycle

$$d\mathbf{V}(t) = \mathbf{V}(t) \sum_{\ell,m=0}^{4} L^m_\ell d\mathbf{\Lambda}^\ell_m(t).$$

The domain D is the core for all the operators L^m_ℓ, which permits giving a rigorous meaning to the previous equation. Introducing the stochastic flow

$$\mathbf{j}_t(X) = \mathbf{V}(t) X \mathbf{V}(t)^*, \quad (t \ge 0, X \in \mathcal{B}(h)),$$

a formal derivation of the differential equation of the flow, and using the Itô table of Example 2 (Section 3) yields

$$d\mathbf{j}_t(X) = \sum_{\ell,m=0}^{4} \mathbf{j}_t(\vartheta^\ell_m(X)) d\mathbf{\Lambda}^\ell_m(t), \quad j_0(X) = X,$$

where the structure maps are given by

$$\vartheta^0_0(X) = L^0_0 X + X L^{0\,*}_0 + \sum_{k=1}^{4} L^0_k X L^{0\,*}_k \tag{4.26}$$

$$\vartheta^\ell_m(X) = L^\ell_m X + X L^{\ell\,*}_m, \text{ if } \sup(\ell, m) > 0. \tag{4.27}$$

Therefore, besides the equality $\theta^0_0 = \vartheta^0_0$, the structure maps of j_t and $\mathbf{j}_t$ do not coincide. However, the semigroup given by $X \mapsto \mathbb{E}(\mathbf{j}_t(X))$ has the same generator $\mathcal{L}(X) = \theta^0_0(X)$ as $\mathcal{T}_t(X)$, $(X \in \mathcal{B}(h))$. Thus $\mathbf{j}_t$ may be considered as a dilation of the given quantum dynamical semigroup. However, the flow j_t has an intrinsically richer structure, as we will see in the next section.

5 Unraveling: classical stochastic processes related to quantum flows

A quantum Markov flow when restricted to invariant abelian subalgebras gives raise to different classical stochastic processes. We will illustrate this with the quantum optics flow introduced in the previous section.

We start by describing some invariance properties of $\mathcal{L}_0$. The same invariance properties of the corresponding free quantum dynamical semigroup can be easily proved rigorously using an integral equation satisfied by this semigroup; see, for instance, [15].

5.1 The algebra generated by the number operator

Let us consider the abelian W^*-subalgebra of $\mathcal{B}(h)$ generated by the operator N, that is

$$\mathcal{A}_N = \left\{ \varphi(N) \mid \varphi \in l^\infty(\mathbb{N}; \mathbb{C}) \right\}.$$

Proposition 5.1. *The restriction of the free generator $\mathcal{L}_0$ to the abelian subalgebra $\mathcal{A}_N$ coincides with the generator of a classical birth and death process.*

Proof. A straightforward computation using the well-known commutation relations yields

$$\mathcal{L}_0\left(\varphi(N)\right) = \mu^2 N\left(\varphi(N-1) - \varphi(N)\right) + \lambda^2 (N+1)\left(\varphi(N+1) - \varphi(N)\right),$$

for all appropriate functions φ. Therefore, the restriction of $\mathcal{L}_0$ to $\mathcal{A}_N$ agrees with the infinitesimal generator of a classical *birth-and-death* process with birth rates $\left(\lambda^2(k+1)\right)_{k\geq 0}$ and death rates $\left(\mu^2 k\right)_{k\geq 0}$. □

The complete quantum evolution is obtained by adding to the free Liouvillian $\mathcal{L}_0$ the bounded perturbation $\mathcal{L}_{ga}$ (called the "gain operator") given by

$$\mathcal{L}_{ga}(X) = R^2\left(\varphi_1(N)X\varphi_1(N) + \varphi_2(N)S^*XS\varphi_2(N) - X\right).$$

The bounded operator $\mathcal{L}_{ga}$ on $\mathcal{B}(h)$ is the infinitesimal generator of a norm continuous, identity preserving and completely positive semigroup on $\mathcal{B}(h)$.

In [16] it has been shown that the restriction of $\mathcal{L}$ to the von Neumann algebra $\mathcal{A}_N$ generated by the number operator agrees with the infinitesimal generator of a birth and death process. In fact it is easy to see that the restriction to $\mathcal{A}_N$ of the operator $\mathcal{L}_{ga}$ agrees with the infinitesimal generator of a pure birth process with birth rates $(\sin^2(\phi\sqrt{k+1}))_{k\geq 0}$.

Proposition 5.2. *The restriction of the whole evolution to the algebra generated by the number operator corresponds to a birth and death process with birth rates*

$$\lambda_k = \lambda^2(k+1) + R^2\sin^2(\phi\sqrt{k+1}) \text{ if } k \geq 0, \tag{5.1}$$

and death rates

$$\mu_k = \mu^2 k \text{ if } k \geq 0. \tag{5.2}$$

5.2 Position, momentum and their algebras

The *momentum (or electric field) operator* is defined by

$$D(p) = D(a) = D(a^\dagger), \qquad p = \frac{i}{\sqrt{2}}\left(a^\dagger - a\right). \tag{5.3}$$

The *position (or magnetic field) operator* is given by

$$D(q) = D(a) = D(a^\dagger), \qquad q = \frac{1}{\sqrt{2}}\left(a^\dagger + a\right). \tag{5.4}$$

The above operators satisfy the canonical commutation relation (CCR),

$$[q, p] = qp - pq = iI,$$

(since we take the Planck constant $\hbar = 1$ as a convention).

All the above computations have been done in the Heisenberg representation of the CCR over $\mathbb{C}$, which is equivalent to the representation due to Schrödinger. More precisely, the equivalence is obtained through a unitary operator $\mathcal{U}$ defined as follows. Denote $(H_n)_{n\geq 0}$ the orthonormal sequence of the Hermite polynomials in $L^2(\mathbf{R}; \pi^{-1/4}\exp(-x^2/2)dx)$ and define

$$\mathcal{U} : l^2(\mathbb{N}) \to L^2(\mathbf{R}; \pi^{-1/4}\exp(-x^2/2)dx), \qquad \mathcal{U}e_n = H_n.$$

We refer to Meyer's book [34] for more details on this subject. Here we shall use the following "conversion table."

Heisenberg	Schrödinger
p	$-i\frac{d}{dx}$
q	x
N	$\frac{1}{2}\left(-\frac{d^2}{dx^2} + x^2 - 1\right)$

Let $\mathcal{A}_q$, $\mathcal{A}_p$ be the following abelian von Neumann subalgebras of $\mathcal{B}(h)$

$$\begin{aligned} \mathcal{A}_q &= \left\{ \varphi(q) \mid \varphi \in L^\infty(\mathbb{R}; \mathbb{C}) \right\}, \\ \mathcal{A}_p &= \left\{ \varphi(p) \mid \varphi \in L^\infty(\mathbb{R}; \mathbb{C}) \right\}, \end{aligned}$$

where $L^\infty(\mathbb{R}; \mathbb{C})$ denotes the vector space of all complex-valued bounded measurable functions on $\mathbb{R}$.

Now we will compute, formally, the action of $\mathcal{L}_0$ on the abelian algebra $\mathcal{A}_q$, $\mathcal{A}_p$.

Proposition 5.3. *The restriction of $\mathcal{L}_0$ to the abelian algebra $\mathcal{A}_q$ coincides with the infinitesimal generator of an* Ornstein-Uhlenbeck *process; in particular, this process reduces to a Brownian motion with variance λ^2 if $\lambda = \mu$.*

Proof. We will use the commutation relations

$$[a, \varphi(q)] = \frac{1}{\sqrt{2}}\varphi'(q), \qquad \left[a^\dagger, \varphi(q)\right] = -\frac{1}{\sqrt{2}}\varphi'(q), \tag{5.5}$$

$$[a, \varphi(p)] = \frac{i}{\sqrt{2}}\varphi'(p), \qquad \left[a^\dagger, \varphi(p)\right] = \frac{i}{\sqrt{2}}\varphi'(p). \tag{5.6}$$

which can be easily checked working in the Schrödinger representation.

Let $\varphi \in C_b^2(\mathbb{R}; \mathbb{C})$. Then, identifying $\varphi(q)$ with the corresponding multiplication operator on $L^2(\mathbf{R}; \pi^{-1/4}\exp(-x^2/2)dx)$, we have

$$\begin{aligned}
\mathcal{L}_0(\varphi(q)) &= \frac{\mu^2}{2}\left(-a^\dagger a\varphi(q) + 2a^\dagger\varphi(q)a - \varphi(q)a^\dagger a\right) \\
&+ \frac{\lambda^2}{2}\left(-aa^\dagger\varphi(q) + 2a\varphi(q)a^\dagger - \varphi(q)aa^\dagger\right) \\
&= \frac{\mu^2}{2}\left(-a^\dagger\,[a, \varphi(q)] + \left[a^\dagger, \varphi(q)\right]a\right) \\
&+ \frac{\lambda^2}{2}\left(-a\left[a^\dagger, \varphi(q)\right] + [a, \varphi(q)]\,a^\dagger\right) \\
&= \frac{\lambda^2+\mu^2}{4}\varphi''(q) + \frac{\lambda^2-\mu^2}{2}q\varphi'(q).
\end{aligned}$$

Thus $\mathcal{L}_0$ transforms the multiplication operator by φ into the multiplication operator by the function

$$q \to \frac{\lambda^2+\mu^2}{4}\varphi''(q) + \frac{\lambda^2-\mu^2}{2}q\varphi'(q),$$

i.e., the restriction of $\mathcal{L}_0$ to the abelian algebra $\mathcal{A}_q$ agrees with the infinitesimal generator of an *Ornstein–Uhlenbeck* process (a *Brownian motion* with variance λ^2 if $\lambda = \mu$).□

A similar computation shows that, for all $\varphi \in C_b^2(\mathbb{R}; \mathbb{C})$, we have

$$\mathcal{L}_0(\varphi(p)) = \frac{\lambda^2+\mu^2}{4}\varphi''(p) + \frac{\lambda^2-\mu^2}{2}p\varphi'(p).$$

Proposition 5.4. *The restriction of $\mathcal{L}_0$ to the abelian algebra $\mathcal{A}_p$ coincides with the infinitesimal generator of an* Ornstein-Uhlenbeck *process (a* Brownian motion *with variance λ^2 if $\lambda = \mu$).*

Remark. The abelian algebras $\mathcal{A}_q$, $\mathcal{A}_p$ generated by the position (or magnetic field) and momentum (or electric field) operator *are not invariant* under the action of $\mathcal{L}_{ga}$.

Consider, for example, the algebra $\mathcal{A}_q$. This is the maximal abelian W^*-subalgebra of $\mathcal{B}(h)$ generated by the operators $W(iv)$ with $v \in \mathbb{R}$, where the notation W is used for the *Weyl operators* defined by

$$W(z) = \exp(za^\dagger - \bar{z}a), \quad (z \in \mathbb{C}).$$

These operators satisfy the (CCR)

$$W(z)W(u) = W(z+u)\exp(i\Im(\bar{u}z)). \tag{5.7}$$

A straightforward computation using the above relation shows that for all $u, v \in \mathbf{R} - \{0\}$, we have

$$W(iv)\mathcal{L}_{ga}(W(iu))W(-iv) \neq \mathcal{L}_{ga}(W(iu)).$$

This shows that the operators $\mathcal{L}_{ga}(W(iu))$ and $W(iv)$ do not commute. Therefore the operator $\mathcal{L}_{ga}(W(iu))$ does not belong to the algebra $\mathcal{A}_q$, which is the maximal abelian W^*-subalgebra of $\mathcal{B}(h)$ generated by the operators $W(iv)$ with $v \in \mathbb{R}$. The same computation, with Weyl operators of the form $W(-u)$ with $u \in \mathbf{R}$, shows that the algebra $\mathcal{A}_p$ is not invariant for $\mathcal{L}_{ga}$ either.

As a result, the total evolution ruled by the generator $\mathcal{L}$ does not leave invariant the algebras generated by position and momentum operators. Thus, the semigroup $\mathcal{T}$ restricted to those algebras cannot produce a commutative Markov semigroup.

Final comments

The concept of complete positive maps seems to be well adapted to extending notions related to Markov processes into a noncommutative framework. On the other hand, that concept seems to be the natural mathematical expression of Schrödinger's probabilistic interpretation of quantum mechanics equations. In a sense, his idea of a two-fold process obtained from a forward and a backward equation is reflected by $(V(t), V(t)^*)$ determined by the cocycle equation.

Classical dilations of QDS, which are connected with stochastic Schrödinger equations, have been extensively used by numerous authors, among them, Barchielli, Holevo, Gisin and Percival. For a good account on the state of this theory, the reader is referred to the book of Percival [41].

Open quantum systems represent a privileged terrain of research for probabilists interested in applications to physics.

Another approach to quantum mechanics well known to classical probabilists is the one started by Nelson in [35], [36]. Indeed, his approach uses the classical theory of stochastic processes and was inspired by a fundamental idea going back to Einstein who wanted to establish quantum mechanics within the formalism of classical mechanics.

Apart from some criticism from physicists themselves, the main mathematical problem of stochastic mechanics is connected with the narrow concept of randomness assumed to describe quantum mechanics. More precisely, it is always possible to use a classical probability model to describe the evolution of one fixed observable X of a quantum system, so one can construct a probability space $(\Omega_X, \mathcal{F}_X, \mathbb{P}_X)$, depending on X, and a stochastic process $(X_t)_{t\geq 0}$ to describe the evolution of that quantum observable. Unfortunately, there is little chance that this setting could also serve to describe the evolution of another observable Y, especially if it does not commute with X. In addition, the idea of using classical diffusions to model position and momentum evolution as an approach to quantum mechanics involves a secondary mathematical difficulty: that of making sense of solutions of nonlinear martingale problems with singular coefficients (see for instance [43]).

However, there is a better way to rescue the main idea of stochastic mechanics. That is, the equations of stochastic mechanics may be understood as a *simulation* of a quantum stochastic differential equation. Indeed, given a quantum Markov process j or the corresponding quantum dynamical semigroup $\mathcal{T}$, we can state the following problem. Let a pure state $\rho = |\psi\rangle\langle\psi|$ and an observable X be fixed. The problem consists in studying whether there exists a classical Markov semigroup $(P_t)_{t\geq 0}$ and a Markov process $(X_t)_{t\geq 0}$ defined over a probability space $(\Omega, \mathcal{F}, \mathbb{P})$ such that

$$P_t(f) = \mathrm{tr}\,\rho\mathcal{T}_t(f(X)) = \mathbb{E}(f(X_t)),$$

for all $t \geq 0$ and any bounded measurable function f.

Acknowledgements. The author is gratefully indebted to the organizers of the workshop in Stochastic Analysis and Mathematical Physics held in Lisbon for their kind hospitality and stimulating scientific discussions. Ana Bela Cruzeiro and Jean-Claude Zambrini have generously dedicated much effort to create a special working atmosphere around them in which science combines with friendship.

References

[1] L.Accardi, On the quantum Feynman–Kac formula, *Rend. Sem. Matematico e Fisico di Milano,* XLVIII, 1978.

[2] L. Accardi, F. Fagnola, and J. Quaegebeur, A Representation Free Quantum Stochastic Calculus, *Journal of Functional Analysis,* **104**:1 (1992), 149–197.

[3] L. Accardi and F. Fagnola, Stochastic integration. In *Quantum probability and Application III (Proceedings Oberwolfach, 1987)*, Vol. 1303, Lecture Notes in Math., Springer, 1988, pp. 6–19.

[4] L. Accardi and J. Quaegebeur, The Itô algebra of quantum Gaussian fields, *J. Funct. Anal.,* **85** (1989), 212–229.

[5] S. Albeverio, K. Yasue, and J. C. Zambrini, *Ann. Inst. H. Poincaré (Phys. Théor.),* **49**:3 (1989), 259.

[6] W.B. Arveson, Subalgebras of C^*-algebras, *Acta Math.,* **123** (1969), 141–224.

[7] V.P. Belavkin, A quantum non adapted Itô formula and non stationary evolution in Fock scale, in *Quantum Probability and Related Topics*, World Scientific, vol.VI, (1992), pp. 137–180.

[8] V.P. Belavkin, Nondemolition measurement and nonlinear filtering of quantum stochastic processes, Lect. Notes in Control and Information Sci., Springer, 1988, pp. 245–266.

[9] V.P. Belavkin, Quantum stochastic calculus and quantum nonlinear filtering, *J. of Multivariate Analysis*, **42**:2 (1992), 171–201.

[10] V.P. Belavkin and C. Bendjaballah, Continuous measurements of quantum phase, *Quantum Opt.*, **6** (1994), 169–186.

[11] A.B. Cruzeiro, W. Liming, and J.-C. Zambrini, Bernstein Processes associated with a Markov process, *Proceedings ANESTOC'98*, Birkhäuser, 2000.

[12] P.A.M. Dirac, *Proc. Royal Soc. London A*, **114** (1927), 243.

[13] E.B. Davies, Quantum dynamical semigroups and the neutron diffusion equation, *Rep. Math. Phys.*, **11** (1977), 169–188.

[14] F. Fagnola, On quantum stochastic differential equations with unbounded coefficients, *Prob. Th. and Rel. Fields*, **56** (1990), 501–516.

[15] F. Fagnola, Unitarity of solutions to quantum stochastic differential equations and conservativity of the associated semigroups, *Quantum Probability and Related Topics*, **VII** (1992), 139–148.

[16] F. Fagnola, R. Rebolledo, C. Saavedra, Quantum flows associated to a class of laser master equations, *J. Math. Phys.*, **35**:1 (1994), 1–12.

[17] F. Fagnola, R. Rebolledo, C. Saavedra, Reduction of noise by squeezed vacuum, *Proc. ANESTOC '96*, World Scientific, 1998, pp. 53–63.

[18] F. Fagnola, R. Rebolledo, An ergodic theorem in quantum optics, Proc. Conference in honour of A. Frigerio, Editrice Universitaria Udinese, 1996, 73–86.

[19] F. Fagnola, R. Rebolledo, Sur l'approche de l'équilibre au moyen des flots quantiques, *C.R. Acad. Sci. Paris, sér. I*, **321** (1995), 473–476.

[20] F. Fagnola and R. Rebolledo, The approach to equilibrium of a class of quantum dynamical semigroups, *Inf. Dim. Anal. Q. Prob. and Rel. Topics*, **1**:4 (1998), 1–12.

[21] F. Fagnola, R. Rebolledo, A probabilistic view on stochastic differential equations derived from Quantum Optics, *Aportaciones Matemáticas*, Soc. Matem. Mexicana, **14**(1998), 193–214.

[22] A. Frigerio, Quantum dynamical semigroups and approach to equilibrium, *Lett. Math. Phys.* **2**(1977), 79–87.

[23] B. Jamison, *Z. Wahrschernlichkeit V. Gebiete* **30** (1974), p. 65.

[24] A. Kossakowski, V. Gorini and E.C.G. Sudarshan, Completely positive dynamical semigroups of n-level systems, *J. Math. Phys.*, **17**(1976), 821–825.

[25] H. Haken, *Handbuch der Physik*, volume XXV/2c. Springer, Berlin, 1969.

[26] E. Hille, *Functional Analysis and Semigroups*, AMS, 1957.

[27] A.S. Holevo, *Probabilistic and Statistical aspects of Quantum Theory*, North-Holland, 1982.

[28] R.L. Hudson and K.R. Parthasarathy, Quantum Itô's formula and stochastic evolutions, *Comm. Math. Phys.*, 93 (1984), 301–323.

[29] E.T. Jaynes and F.W. Cummings, *Proc.IEEE*, **51**:89, 1963.

[30] K. Kraus, General states changes in Quantum Theory, *Ann. Phys.*, **64** (1970), 311–335.

[31] G. Lindblad. On the generators of quantum dynamical semigroups. *Commun. Math. Phys.*, 48:119–130, 1976.

[32] W.H. Louisell, *Quantum Statistical Properties of Radiation*, John Wiley, N.Y., 1973.

[33] L.A. Lugiato, M.O. Scully, and H. Walther, *Phys. Rev. A*, 36(1987), 740.

[34] P.-A. Meyer, *Quantum Probability for Probabilists*, LNM 1538, Springer, Heidelberg, 1993.

[35] E. Nelson, *Dynamical Theories of Brownian Motion*, Princeton University Press, 1967.

[36] E Nelson, *Quantum Fluctuations*, Princeton University Press, 1985.

[37] M. Orszag, *Quantum Optics*, Springer, 2000.

[38] K.R. Parthasarathy and K.B. Sinha, Markov chains as Evans-Hudson diffusions in Fock space, *Séminaire de Probabilités*, LNM, Springer XXIV(1426), 1988–1989, pp. 362–369.

[39] K. R. Parthasarathy, *An Introduction to Quantum Stochastic Calculus*, Vol. 85, Monographs in Mathematics, Birkhaüser, Basel, 1992.

[40] A. Pazy, *Semigroup of Linear Operators and Applications to Partial Differential Equations*, Springer, 1975.

[41] I. Percival, *Quantum State Diffusion*, Cambridge University Press, 1999.

[42] R. Rebolledo, J.C. Retamal, C. Saavedra, Diffusion processes associated to a laser model, *J. Math. Phys.*, **33**(1992), 826–831.

[43] R. Rebolledo, The Role of weak topologies in Stochastic Mechanics, Ed. Bernoulli, Proc. IV CLAPEM, Acta Científica Venezolana, Caracas, no. 3 (1992), 43–60.

[44] R. Rebolledo, Sur les semigroupes dynamiques quantiques, *Ann. Math. Blaise Pascal*, **3**:1(1996), 125–142.

[45] M. Sargent, M.O. Scully, and W.E. Lamb, *Laser Physics*, Addison-Wesley, 1974.

[46] M.O. Scully and W.E. Lamb, Quantum Theory of an Optical Maser. I. General Theory, *Phys. Rev.*, **159**(1967), 208–226.

[47] W.F. Stinespring, Positive functions on C^*-algebras, Proc. Am. Math. Soc., **6** (1955), 211–216.

[48] L. Susskind and J. Glogower, *Physics*, 1:49 (1964).

[49] E. Schrödinger, *Ann. Inst. H. Poincaré*, **2** (1932), 269.

[50] J. C. Zambrini, *J. Math. Phys.* **27**:9 (1986), 2307.

R. Rebolledo
Facultad de Matemáticas
Universidad Católica de Chile
Casilla 306 Santiago 22, Chile
e-mail:rrebolle@mat.puc.cl

Properties of Measure-preserving Shifts on the Wiener Space

A.S. Üstünel

ABSTRACT Let (W, H, μ) be an abstract Wiener space and let $T : W \to W$ be a measurable map of the form $T = I_W + u$, where u is a Wiener functional with values in the Cameron–Martin space H. Assume that μ is invariant under T and that almost surely, the map $h \to u(w + h)$ is Fréchet differentiable on H. Then we prove some pointwise properties of T; namely, that T has a universally measurable right inverse that is absolutely continuous, and the set of nondegeneracy of T has full μ-measure. Finally, in the case $E[|\Lambda|] \leq 1$ (see the introduction), we show that T, in fact, is almost surely a bijection.

1 Introduction and preliminaries

Let (W, H, μ) be an abstract Wiener space, i.e., W is a separable Banach space, H is a separable Hilbert space, and there is a continuous and dense injection from H into W. μ denotes the Gauss measure on W whose reproducing kernel Hilbert space is H, which is also called the Cameron–Martin space. The problem we consider in this work is the following: suppose that there is a measurable mapping $T : W \to W$ of the form $T = I_W + u$, where I_W is the identity map of W and $u : W \to H$ is measurable. Assume that the image of the Wiener measure μ under T, denoted by $T^*\mu$, is equal to μ. What can we say about the pointwise properties of T? Of course, in this generality, it is almost impossible to progress; hence we have to make some reasonable hypothesis about u so that the results hold true for T in the change of variables formula or Sard's theorem (cf. [6]). Let us first explain some notation and recall some preliminary results.

We denote by ∇ the Sobolev derivative defined on W which is an $L^p(\mu)$-extension of the Gateaux derivative in the Cameron–Martin space direction (cf. [4]). For a separable Hilbert space Ξ, $p > 1$ and $\beta \in \mathbb{R}$, $\mathbb{D}_{p,\beta}(\Xi)$ denotes the Sobolev space of Ξ-valued Wiener functionals such that

$$\left\|(I + L)^{\beta/2}\xi\right\|_{L^p(\mu,\Xi)} < \infty,$$

where L denotes the Ornsterin–Uhlenbeck operator (cf. [4]). For the case $\beta \in \mathbb{N}$, due to Meyer inequalities (cf. [4] for example), the above norm is equivalent to the one defined as

$$\sum_{i=0}^{\beta} \left\|\nabla^i \xi\right\|_{L^p(\mu,\Xi\otimes H^{\otimes i})}$$

where $\otimes$ denotes the completed Hilbert–Schmidt tensor product and $H^{\otimes n}$ is defined inductively as $H^{\otimes(n-1)} \otimes H$. Consequently, we can extend the operator ∇ to all the Sobolev scale, and this gives a continuous and linear map $\nabla : \mathbb{D}_{p,\beta}(\Xi) \to \mathbb{D}_{p,\beta-1}(\Xi \otimes H)$ for any $p > 1$ and $\beta \in \mathbb{R}$. Consequently, its adjoint operator, denoted as δ, extends continuously from $\mathbb{D}_{p,\beta}(\Xi \otimes H)$ into $\mathbb{D}_{p,\beta-1}(\Xi)$, for any $p > 1$, $\beta \in \mathbb{R}$.

A measurable map $u : W \to H$ is called $H - C^1$ if outside a μ-negligeable set N, the map $h \to u(w+h)$ is Fréchet differentiable on H for any $w \in N^c$. Note that in this case N is an H-invariant set, hence of zero $\mathbb{D}_{p,1}$ capacity for any $p > 1$; moreover ∇u and δu are well-defined (see, e.g., [3, 6]). We have then the following result (see [6]):

Theorem 1. *Assume that $T = I_W + u$, where u is a measurable, H-valued, $H - C^1$-map. Define the set*

$$M = \{w \in W : \det_2(I_H + \nabla u(w)) \neq 0\},$$

where $\det_2(I_H + \nabla u(w))$ denotes the modified Carleman–Fredholm determinant of the operator $I_H + \nabla u(w)$ (see [2]), which is a linear and continuous operator on H, and $\nabla u(w)$ is Hilbert–Schmidt.

- *Then μ-almost surely, the cardinal of the set $T^{-1}\{w\} \cap M$, is at most countably infinite. Besides for any f, g, continuous and bounded positive functions on W, we have*

$$E\left[f \circ T\, g\, |\Lambda|\right] = E\left[f \sum_{y \in T^{-1}\{w\} \cap M} g(y)\right], \tag{1.1}$$

  where

$$\Lambda = \det_2(I_H + \nabla u) \exp\left\{-\delta u - 1/2|u|_H^2\right\}.$$

  This implies also that $T^*(\mu|_M)$ is absolutely continuous with respect to μ.

- *We have, for any $A \in \mathcal{B}(W)$, that $T(A)$ is universally measurable, and*

$$\mu(T(A)) \leq \int_A |\Lambda|\, d\mu. \tag{1.2}$$

  This implies in particular that $\mu(T(M^c)) = 0$, consequently, at the right hand side of the equality (1.1), one can replace the set $T^{-1}\{w\} \cap M$ by $T^{-1}\{w\}$.

2 Pointwise properties of T

In this section, we assume as in Theorem 1, that $u : W \to H$ is $H - C^1$. We have

Proposition 1. *Assume that $T^*\mu = \mu$. Then $T^*(\mu|_M)$ is equivalent to the Wiener measure μ, where $T^*(\mu|_M)$ denotes the image under T of the restriction of μ to the set $M = \{\det_2(I_H + \nabla u) \neq 0\}$.*

Proof. Let us denote by $N(w)$ the random variable that is equal to the cardinal of the set $T^{-1}\{w\} \cap M$. From Theorem 1, this cardinal is equal almost surely to that of $T^{-1}\{w\}$. Then, for any $f \in C_b(W)$, positive, we have from the same theorem

$$
\begin{aligned}
E[f \circ T\, |\Lambda|] &= E[f N] \\
&= E[f \circ T\, N \circ T] \\
&\geq E[f \circ T] \\
&= E[f]
\end{aligned}
$$

since μ is T-invariant and since $N \circ T \geq 1$ almost surely. Hence μ is absolutely continuous with respect to $T^*(\mu|_M)$. The fact that $T^*(\mu|_M) \ll \mu$ is true for general $H - C^1$-shifts, it also follows trivially from $T^*\mu = \mu$. □

Proposition 2. *Under the hypothesis of Proposition 1, T has a universally measurable right inverse.*

Proof. Since, from Proposition 1, $T^*(\mu|_M)$ is equivalent to μ, the Radon–Nikodym derivative of the former with respect to latter is different than zero μ almost surely. Note that this derivative is equal to

$$
\frac{dT^*(\mu|_M)}{d\mu} = \sum_{y \in T^{-1}\{w\}} \frac{1}{|\Lambda(y)|}\,.
$$

Hence almost surely $T^{-1}\{w\} \neq \emptyset$. Let us define the multivalued map K as

$$
K(w) = \{h \in H : h + u(w + h) = 0\}\,.
$$

From the above observation, it follows that $K(w) \neq \emptyset$ almost surely, and moreover its graph is $\mathcal{B}(H \times W)$-measurable. It follows from von Neumann's selection theorem that there exists a universally measurable H-valued map v on W such that $v(w) \in K(w)$ almost surely (see [1]). It is then immediate to see that $S = I_W + v$ is a right inverse. □

Proposition 3. *The image of μ under the right inverse S, i.e., $S^*\mu$, is absolutely continuous with respect to μ.*

Proof. Let D be the set $\{w \in W : T(S(w)) = w\}$ from Proposition 2, $\mu(D) = 1$. Moreover, for any $A \in \mathcal{B}(W)$, if $w \in D$ and $S(w) \in A$, then $w = T(S(w)) \in T(A)$, and hence from (1.2)

$$
\begin{aligned}
\mu(D \cap S^{-1}(A)) &= \mu(S^{-1}(A)) \\
&\leq \mu(D \cap T(A)) = \mu(T(A)) \\
&\leq E[\mathbf{1}_A |\Lambda|]\,.
\end{aligned}
$$

Consequently $S^*(\mu|_D) = S^*\mu \ll \mu$. □

We have finally

Theorem 2. *The set of nondegeneracy M has full μ-measure.*

Proof. Suppose that $\mu(M^c) \neq 0$ and define

$$\nu = \frac{1}{\mu(M^c)}\mu|_{M^c}\,.$$

Then

$$T^*\nu(T(M^c)) = \nu(T^{-1}(TM^c)) \geq \nu(M^c) = 1\,.$$

However $\mu(T(M^c)) = 0$ from the inequality (1.2). Consequently, $T^*\nu$ and μ are mutually singular. On the other hand,

$$\begin{aligned} T^*\nu &= \frac{1}{\mu(M^c)}T^*(\mu|_{M^c}) \\ &= \frac{1}{\mu(M^c)}(\mu - T^*(\mu|_M))\,, \end{aligned}$$

and this last relation implies that $T^*\nu \ll \mu$, which gives a contradiction to the assumption $\mu(M^c) \neq 0$. □

Corollary 1. *We have*

$$\sum_{y \in T^{-1}\{w\}} \frac{1}{|\Lambda(y)|} = 1$$

μ-almost surely.

Proof. Since $\mu(M^c) = 0$, $|\Lambda| \neq 0$ μ-almost surely. Hence, for any $f \in C_b(W)$,

$$\begin{aligned} E[f] &= E[f \circ T] \\ &= E\left[f \circ T \frac{|\Lambda|}{|\Lambda|}\right] \\ &= E\left[f \sum_{y \in T^{-1}\{w\}} \frac{1}{|\Lambda(y)|}\right] \end{aligned}$$

and the proof follows. □

Theorem 3. *Assume that u is $H - C^1$, $T^*\mu = \mu$ and that $E[|\Lambda|] \leq 1$. Then T is almost surely a bijection and $|\Lambda| = 1$.*

Proof. We have, for any positive, measurable f on W,

$$E[f \circ T\,|\Lambda|] = E[f\,N] \geq E[f]\,.$$

Hence $N \geq 1$ almost surely. Moreover,

$$1 \geq E[|\Lambda|] = E[N] \geq 1\,,$$

and hence $E[N] = 1$, which implies that $N = 1$ almost surely, and this implies that $E[|\Lambda|] = 1$. Let U be defined on the set $D' = \{w \in W : N(w) = 1\}$ as $U(Ty) = y$, since $D' \subset T(W)$, U is well-defined on D' and $\mu(D') = 1$. Furthermore, for any $w \in D \cap D' \cap S^{-1}(D')$, we have

$$U \circ T(Sw) = Sw.$$

Also, since $w \in D$, $T(Sw) = w$, hence $U(T \circ S(w)) = U(w)$, consequently $U = S$ almost surely. We have trivially that $S^*\mu = \mu$, and it follows from Corollary 1 that $|\Lambda| = 1$ almost surely. □

References

[1] C. Castaing and M. Valadier, *Convex Analysis and Measurable Multifunctions*, Lecture Notes in Math. 580, Springer, 1977.

[2] N. Dunford and J.T. Schwartz, *Linear Operators* **2**, Interscience, 1963.

[3] S. Kusuoka, The nonlinear transformation of a Gaussian measure on Banach space and its absolute continuity, *J. Fac. Sci., Tokyo Univ., Sect. 1.A.*, **29** (1982), 567–590.

[4] A.S. Üstünel, *Introduction to Analysis on Wiener Space*, Lecture Notes in Math., Vol. 1610, Springer, 1995.

[5] A.S. Üstünel and M. Zakai, Random rotations of the Wiener path, *Prob. Theory Rel. Fields,* **103** (1995), 409–430.

[6] A.S. Üstünel and M. Zakai, *Transformation of Measure on Wiener Space*, Springer Verlag, 1999.

A. S. Üstünel,
ENST, Dépt. Réseaux,
46 Rue Barrault, 75013 Paris, France
ustunel@enst.fr

Martingale and Markov Uniqueness of Infinite Dimensional Nelson Diffusions

Liming Wu

1 Introduction

In this paper we study the uniqueness of a solution of the martingale problem associated with a generalized Schrödinger operator or generator of a Nelson diffusion on a general Polish space E, given by

$$\mathcal{L}^{\phi} f = \mathcal{L} f + 2\frac{\Gamma(f, \phi)}{\phi}, \quad \forall f \in \mathcal{D} \tag{1.1}$$

where $\mathcal{D}$ is some space of testfunctions, $\mathcal{L}$ is the generator of some E-valued Markov process (say, the non-perturbed or free process), $\Gamma(\cdot, \cdot)$ is the associated *square-field* operator, and ϕ is some wave function belonging to the Dirichlet space associated with $(\mathcal{L}, \mu)$. To make it clear, let us begin by describing the free process.

1.1 Assumptions on the free process and on $\mathcal{D}$

Let E be a Polish space with Borel field $\mathcal{B}$, μ a σ-finite measure on $(E, \mathcal{B})$ charging all open subsets of E. First, our free process will be a *conservative continuous* Markov–Hunt process with states space E, realized canonically on $\big(\mathbf{C}(\mathbb{R}^+, E)$, (X_t), $(\mathbb{P}_x)_{x\in E}\big)$ (where $\mathbf{C}(\mathbb{R}^+, E)$ is the space of continuous applications from $\mathbb{R}^+$ to E, (X_t) is the system of coordinates, $\mathbb{P}_x$ is the law of our free process from $x \in E$), such that its transition semigroup $(P_t(x, dy))$ is symmetric on $L^2(E, \mu)$, i.e.,

$$\langle P_t f, g\rangle_\mu = \langle f, P_t g\rangle_\mu := \int_E f P_t g d\mu, \quad \forall f, g \in L^2(E, \mu). \tag{1.2}$$

Let $(\mathcal{L}, \mathbf{D}_2(\mathcal{L}))$ be the generator of (P_t) in $L^2(E, \mu)$, and

$$\mathcal{E}(f, g) = \langle \sqrt{-\mathcal{L}} f, \sqrt{-\mathcal{L}} g\rangle_\mu, \quad \forall f, g \in \mathbf{D}(\sqrt{-\mathcal{L}}) := \mathbf{D}(\mathcal{E}) \tag{1.3}$$

its associated Dirichlet form. $\mathbf{D}(\mathcal{E})$ is a Hilbert space with inner product

$$\mathcal{E}_1(f, g) = \langle f, g\rangle_\mu + \mathcal{E}(f, g), \quad \forall f, g \in \mathbf{D}(\mathcal{E}).$$

Throughout this paper we assume that our space $\mathcal{D}$ of test functions verifies

(C.i) $\mathcal{D}$ is a countably generated subalgebra of the space $C_b(E)$ of real bounded continuous functions on E;

(C.ii) $\mathcal{D}$ separates the points of E (i.e., $\forall x \in E, \exists f \in \mathcal{D} : f(x) \neq 0$);

(C.iii) $\mathcal{D} \subset \mathbf{D}_2(\mathcal{L})$ and $\mathcal{L}u \in L^\infty(\mu)$ for all $u \in \mathcal{D}$;

(C.iv) $\mathcal{D}$ is a form core of $(\mathcal{E}, \mathbf{D}(\mathcal{E}))$.

Consider the square field operator

$$\Gamma(f) = \Gamma(f, f), \ \Gamma(f, g) := \frac{1}{2}\left(\mathcal{L}(fg) - f\mathcal{L}g - g\mathcal{L}f\right), \ \forall f, g \in \mathcal{D}. \quad (1.4)$$

Note that $\Gamma(f) \in L^\infty(\mu)$ for all $f \in \mathcal{D}$ by (C.i) and (C.iii). It is well known that $\Gamma(\cdot, \cdot)$ can be extended as a definite nonnegative and symmetric bilinear form over the whole $\mathbf{D}(\mathcal{E}) \times \mathbf{D}(\mathcal{E})$ such that

$$\mathcal{E}(f, g) = \int_E \Gamma(f, g) d\mu, \quad \forall f, g \in \mathbf{D}(\mathcal{E}). \quad (1.5)$$

Now, by the theory of Dirichlet forms ([8] or [6]), the Hunt property and the path continuity are equivalent, respectively, to:

(C.v) the capacity associated with $\mathcal{E}_1$ is tight;

(C.vi) Γ is a derivation: $\forall f_i, g \in \mathbf{D}(\mathcal{E}), i = 1, \ldots, n$ and $F \in C^1(\mathbb{R}^n)$ with continuous bounded derivatives, $F(f_1, \ldots, f_n) \in \mathbf{D}(\mathcal{E})$ and

$$\Gamma(F(f_1, \ldots, f_n), g) = \sum_{i=1}^n \partial_i F(f_1, \ldots, f_n) \cdot \Gamma(f_i, g). \quad (1.6)$$

1.2 Pre-Dirichlet form associated with the Nelson generator

Our wave function ϕ satisfies

$$\phi \in \mathbf{D}(\mathcal{E}) \text{ and } \int_E \phi^2 d\mu = 1. \quad (1.7)$$

As $\phi = 0$ possibly and $\Gamma(\phi, \cdot)$ is well defined only μ-a.e., the generator $\mathcal{L}^\phi$ given in (1.1) is very singular. Set $\mu_\phi := \phi^2 d\mu$.

Remark that for any $f \in \mathcal{D}$, $\mathcal{L}f \in L^\infty(\mu)$ and

$$\int_E \left(\frac{\Gamma(\phi, f)}{\phi}\right)^2 d\mu_\phi = \int_E \Gamma(\phi, f)^2 d\mu \leq \|\Gamma(f)\|_{L^\infty(\mu)} \cdot \int_E \Gamma(\phi) d\mu,$$

hence $\mathcal{L}^\phi f \in L^2(\mu_\phi)$. Moreover we can verify easily that

$$\langle -\mathcal{L}^\phi f, g\rangle_{\mu_\phi} = \int_E \Gamma(f, g)\phi^2 d\mu := \mathcal{E}^\phi(f, g), \quad \forall f, g \in \mathcal{D}. \quad (1.8)$$

It is a pre-Dirichlet form associated with the symmetric operator $(\mathcal{L}^\phi, \mathcal{D})$ on $L^2(E, \mu_\phi)$.

In Nelson's stochastic mechanics, the diffusion generated by $\mathcal{L}^\phi$ is used to model the trajectorial movement of the system at the equilibrum measure $\phi^2 \cdot \mu := \mu_\phi$; see Nelson [10], Meyer and Zheng [9]. In current studies this operator is investigated in two particular but important contexts:

(I) $E = \mathbb{R}^d$ (or an open domain), $\mathcal{L} = \Delta/2$ (generator of the Brownian Motion), $\mu(dx) = dx$, and

$$\mathcal{L}^\phi f = \frac{1}{2}\Delta f + \frac{1}{\phi}\nabla\phi \cdot \nabla f, \ \forall f \in C_0^\infty(E).$$

(II) E is an infinite dimensional vector space, $\mathcal{L}$ is the generator of the (generalized) Ornstein–Uhlenbeck process, μ is a gaussian measure on E, and

$$\mathcal{L}^\phi f = \mathcal{L} f + \frac{1}{\phi}\langle \nabla\phi, \nabla f\rangle_H$$

where ∇ is the differential operator of Malliavin, H is an appropriate Cameron–Martin subspace.

The case (I) corresponds to quantum mechanics, yet (II) does to quantum fields.

1.3 About the existence

For the existence of the diffusion generated by $\mathcal{L}^\phi$, we recall

Theorem A (see Meyer and Zheng [9] (1985), Fukushima et al. [6] (Th.6.3.3, 1994) for the basic parts (a) and (b)). *Fix one quasi-continuous version $\tilde{\phi}$ of ϕ and let*

$$\sigma^\phi := \inf\left\{t \geq 0; \tilde{\phi}(X_t) = 0\right\} \tag{1.9}$$

be the first hitting time to the node set $N_\phi = [\tilde{\phi} = 0]$. Let $(M_t(\phi))_{t\geq 0}$ be the martingale in the Fukushima decomposition of $(\tilde{\phi}(X_t) - \tilde{\phi}(X_0))_{t\geq 0}$. Define

$$L_t^\phi = \int_0^t \frac{1}{\tilde{\phi}(X_s)} dM_s(\phi), \quad \forall t < \sigma^\phi \text{ (stochastic integral)} \tag{1.10}$$

$$e(L^\phi)_t := \exp\left(L_t^\phi - \frac{1}{2}\langle L^\phi\rangle_t\right), \quad \forall t < \sigma^\phi \tag{1.11}$$

($\langle L^\phi\rangle$ being the predictable quadratic variational process of L^ϕ). Then

(a) $\left((e(L^\phi)_t 1_{[t<\sigma^\phi]}\right)_{t\geq 0}$ is a $\mathbb{P}_{\mu_\phi}$-martingale.

(b) Let $\mathbb{Q}^\phi$ *is the probability measure on* $\mathbf{C}(\mathbb{R}^+, \mathbf{E})$ *determined by*

$$\mathbb{Q}^\phi \mid_{\mathcal{G}_t^0} = 1_{[t<\sigma^\phi]} e(L^\phi)_t \cdot \mathbb{P}_{\mu_\phi}, \quad \forall t \geq 0 \tag{1.12}$$

where $\mathcal{G}_t^0 = \sigma(X_s; 0 \leq s \leq t)$*; then*

$$\mathbb{Q}^\phi(\sigma^\phi = +\infty) = 1. \tag{1.13}$$

Moreover $((X_t), \mathbb{Q}^\phi)$ *is a conservative Markov process, whose transition semigroup*

$$Q_t^\phi f(x) = \mathbb{E}^{\mathbb{Q}^\phi}(f(X_t) \mid X_0 = x) = \mathbb{E}^x f(X_t) 1_{[t<\sigma^\phi]} e(L^\phi)_t \tag{1.14}$$

is a symmetric Markov semigroup on $L^2(E, \mu_\phi)$*.*

(c) For every $f \in \mathcal{D}$*,*

$$M_t^\phi(f) := f(X_t) - f(X_0) - \int_0^t \mathcal{L}^\phi f(X_s) ds \tag{1.15}$$

is a $L^2(\mathbb{Q}^\phi)$*-martingale. In other words the law of* $(X_t)_{t\geq 0}$ *under* $\mathbb{Q}^\phi$ *is a solution of the martingale problem associated with* $(\mathcal{L}^\phi, \mathcal{D}, \mu_\phi)$*.*

(d) The Dirichlet form associated with (Q_t^ϕ) *in* $L^2(E, \mu_\phi)$ *defined in (1.14) is an extension of* $(\mathcal{E}^\phi, \mathcal{D})$ *given in (1.8). And the generator* $\mathcal{L}_{(p)}^\phi$ *of* (P_t^ϕ) *in* $L^p(E, \mu_\phi)$ *is an extension of* $(\mathcal{L}^\phi, \mathcal{D})$ *in* $L^p(E, \mu_\phi)$ *for every* $1 \leq p \leq 2$*.*

The complementary parts (c) and (d) follow from (a,b), see [21] for their *proofs*.
A great number of works are published about the existence of this singular diffusion besides the two references [9], [6] mentionned above. The reader is refered to Cattiaux and Léonard [3] and [20] for relevant references.

1.4 Several types of uniqueness

The question of existence was thus settled and the more delicate question of uniqueness remained. There are four types of uniqueness in the current studies (which are concentrated to the two important cases (I) and (II) described previously):

1) The essential self-adjointness (in short: *e.s.a.*) **of** $\mathcal{L}$ **or uniqueness in** $L^2(\mu_\phi)$. In the case (I), see Wielens [17] (1985). For the e.s.a. in the infinite dimensional case (II), see Albeverio, Kondratiev, Röckner [1] (1992), the author [18] (1994) and Song [14] (1996).

In the negative direction the author [20], [21] (1999) shows that when ϕ is an excited state of a Schrödinger operator $-\mathcal{L} + V$, $(\mathcal{L}^\phi, \mathcal{D})$ is **not** e.s.a. in general.

2) The essential Markovian self-adjointness (in short: *e.m.s.a.*). This means that there is only one sub-Markov symmetric C_0-semigroup whose generator is an extension of $(\mathcal{L}^\phi, \mathcal{D})$.

This direction attracts much attention in recent studies. In case (I), Takeda [15], [16] (1992, 1996) identified the maximal Dirichlet form extending $(\mathcal{E}^\phi, \mathcal{D})$. Röckner and Zhang [11], [12] (1992, 1994) established the e.m.s.a. of $(\mathcal{L}^\phi, C_0^\infty(\mathbb{R}^d))$ under the only condition $\phi \in H^1(\mathbb{R}^d)$ (or more generally $H^1_{loc}(\mathbb{R}^d)$). See Cattiaux and Fradon [4] for other proofs of this e.m.s.a.

In the infinite dimensional case (II), the e.m.s.a. was at first established by Song [13] (1994), see also Röckner and Zhang [11], [12].

3) Uniqueness in $L^1(\mu_\phi)$. This means that there is only one C_0−semigroup on $L^1(\mu_\phi)$ whose generator is an extension of $(\mathcal{L}^\phi, \mathcal{D})$. It is obviously stronger than the e.m.s.a.

This type of uniqueness is studied by the author in a serie of works [19], [20], [21] (1998, 1999) for the usual Schrödinger operators $-\Delta/2 + V$ or for $(\mathcal{L}^\phi, \mathcal{D})$. One main result in [21] is the following:

Theorem B. *Besides (C.i) → (C.iv) and (1.7), if moreover*

(i) $\mathcal{D}$ is a core for $(\mathcal{L}, \mathbf{D_2}(\mathcal{L}))$ (i.e., $(\mathcal{L}, \mathcal{D})$ is L^2-unique);

(ii) $\phi^2 \in \mathbf{D}(\mathcal{E})$;

then $(\mathcal{L}^\phi, \mathcal{D})$ is unique in $L^1(\mu_\phi)$.

4) Uniqueness of solution of martingale problem associated with $(\mathcal{L}^\phi, \mathcal{D})$. The martingale uniqueness in finite dimensional case (I) is solved by the author in [20] (Lemma 1.7 and Prop. 3.4), by means of the very rich theory in the finite dimensional case, developed in Jacod-Shiryayev [7], but remains fuzzy in the infinite dimensional case (II). Indeed it is quite difficult to raise the question properly in case (II).

Martingale uniqueness in the infinite dimensional setting is the main object of this paper. Our purpose consists of removing condition (ii) in Theorem B and to substitute its L^2-uniqueness assumption (i) by some probabilistic uniqueness one for the non-perturbed $(\mathcal{L}, \mathcal{D})$. Notably we shall show that martingale uniqueness is stronger than Markov uniqueness.

1.5 Organization

In the next section the main results of this paper are stated. We prove them in Sections 3, 4, 5.

2 Main results

2.1 Martingale problem associated with $(\mathcal{L}^\phi, \mathcal{D})$

We will follow Jacod–Shiryaev [7] (with some more complications because of the infinite dimensional nature here), to properly raise the martingale problem.

Let $E_\Delta = E\bigcup\{\Delta\}$, where Δ is an extra point to E, with $\mathcal{B}_\Delta = \sigma(\mathcal{B}, \{\Delta\})$. Consider the space $\Omega^{\mathcal{D}}$ of all applications $\omega : \mathbb{R}^+ \mapsto E_\Delta$ such that

$$\omega(0) \in E,\ u(\omega.) \text{ is continuous on } [0, \zeta),\ \forall u \in \mathcal{D} \text{ and } \omega(t) = \Delta,\ \forall t \geq \zeta$$

where $\zeta(\omega) := \inf\{t \geq 0; \omega(t) = \Delta\}$ is the *life time* defined on $\Omega^{\mathcal{D}}$. Note that $\zeta(\omega) > 0$ over $\Omega^{\mathcal{D}}$. We write simply $\Omega := \Omega^{\mathcal{D}}$ if no confusion is possible.

Let $(X_t(\omega) = \omega(t); t \geq 0)$ be the process of coordinates on Ω and

$$\mathcal{F}_t^0 = \sigma(X_s; 0 \leq s \leq t),\ \mathcal{F}_t := \bigcap_{s>t} \mathcal{F}_s^0, \quad \forall t \geq 0.$$

We can regard $\mathbb{P}_\nu = \int_E \nu(dx)\mathbb{P}_x$, the law of our free Markov process with initial measure ν, as a probability measure on $\Omega \supset \mathbf{C}(\mathbb{R}^+, E)$, satisfying $\mathbb{P}_\nu(\zeta = +\infty) = 1$.

For any probability measure $\mathbb{Q}$ on Ω, let $(\mathcal{F}_t^0)^{\mathbb{Q}}$ be the completion of the Borel σ–field $\mathcal{F}_t^0$ w.r.t. $\mathbb{Q}$ and set

$$\mathcal{F}_t^{\mathbb{Q}} = \bigcap_{s>t} \left(\mathcal{F}_s^0\right)^{\mathbb{Q}}, \quad \mathcal{F}_t^\nu = (\mathcal{F}_t)^{\mathbb{P}_\nu}.$$

(it is well known that $\mathcal{F}_t^\nu = (\mathcal{F}_t^0)^{\mathbb{P}_\nu}$).

Definition 2.1. Let ν be a probability measure on E such that $\nu \ll \mu_\phi$. A probability measure $\mathbb{Q}$ on the space $\Omega = \Omega^{\mathcal{D}}$ defined above is called a solution to the martingale question associated with $(\mathcal{L}^\phi, \mathcal{D}, \nu)$, if

(D.i) (initial condition) $\mathbb{Q}(X_0 \in \cdot) = \nu$;

(D.ii) (marginal absolute continuity) *for each* $t \geq 0$,

$$\mathbb{Q}(X_t \in dx, t < \zeta) \ll \mu_\phi(dx);$$

(D.iii) (martingale solution) *For each* $u \in \mathcal{D}$ *and* $n \geq 1$, $(M^\phi_{\cdot\wedge\tau_n}(u))$ *is a* $\mathbb{Q}$*-local martingale w.r.t. the filtration* $(\mathcal{F}_t^{\mathbb{Q}})$, *where*

$$M_t^\phi(u) := u(X_t) - u(X_0) - \int_0^t \mathcal{L}u(X_s)ds - 2\int_0^t \frac{\Gamma(\phi, u)}{\phi}(X_s)ds \tag{2.1}$$

and

$$\tau_n := \inf\left\{0 \leq t < \zeta \,\middle|\, E_t(\phi) := \int_0^t \frac{\Gamma(\phi, \phi)}{\phi^2}(X_s)ds > n\right\} \tag{2.2}$$

(convention here: $\inf\emptyset = \zeta$*), where* ϕ *and* $\Gamma(\phi,\phi) \ge 0$ *are some fixed Borel* $\mathcal{B}$*-measurable* μ*-version, and the convention* $\frac{0}{0} = 0$ *is used in the definition of* $E_\cdot(\phi)$ *above.*

Remarks (2.i). To justify our definition above, let ϕ' (resp. $\Gamma(\phi,\phi)' \ge 0$, $\Gamma(\phi,u)'$) be another Borel measurable μ-version of ϕ (resp. $\Gamma(\phi,\phi)$, $\Gamma(\phi,u)$); we should verify that $(\mathbb{P}_\nu + \mathbb{Q})$-a.e.,

$$E_t(\phi) = \int_0^t \frac{\Gamma(\phi,\phi)'}{\phi'^2}(X_s)ds \in [0,+\infty],\ \forall t < \zeta; \tag{2.3}$$

$$\int_0^t \left|\frac{\Gamma(\phi,u)}{\phi}\right|(X_s)ds \vee \int_0^t \left|\frac{\Gamma(\phi,u)'}{\phi'}\right|(X_s)ds < +\infty,\ \forall t \le \tau_n; \tag{2.4}$$

$$\int_0^t \frac{\Gamma(\phi,u)}{\phi}(X_s)ds = \int_0^t \frac{\Gamma(\phi,u)'}{\phi'}(X_s)ds,\ \forall t \le \tau_n. \tag{2.5}$$

(2.3) follows from (D.ii) by Fubini's theorem. In Lemma 3.2 we shall prove (2.4) and (2.5) and the fact that (τ_n) are stopping time w.r.t. the filtration $(\mathcal{F}_t)$. Notice that (2.3) (resp. (2.4) and (2.5)) implies that τ_n (resp. $M^\phi_{\tau_n\wedge\cdot}$) does not depend on μ-versions of ϕ, $\Gamma(\phi)$ and $\Gamma(\phi,u)$, and is $\mathbb{Q}$-well defined.

2.2 Main results

Theorem 2.2. *If a probability measure* $\mathbb{Q}$ *on* $\Omega = \Omega^{\mathcal{D}}$ *is a solution to the martingale problem associated with* $(\mathcal{L}^\phi, \mathcal{D}, \nu)$ *where* $\nu \ll \mu_\phi$*, such that for each* $n \in \mathbb{N}$ *and* $T > 0$*,*

$$\mathbb{Q} \ll \mathbb{P}_\nu \quad \text{over } \mathcal{F}_{\tau_n\wedge T} \tag{2.6}$$

then for each $T > 0$*,*

$$\mathbb{Q}|_{\mathcal{F}_T} = \mathbb{Q}^\phi_\nu|_{\mathcal{F}_T} := 1_{[T<\sigma^\phi]} e(L^\phi)_T \cdot \mathbb{P}_\nu \tag{2.7}$$

where the notations of Theorem A are used.

Remarks (2.ii). When $\mathcal{L} = \frac{1}{2}\Delta$, the drift or the forward velocity of $\mathcal{L}^\phi$ is $(\nabla\phi/\phi)(X_t)$ and hence $E_t(\phi)$ is just the kinetic energy of the diffusion associated with $\mathcal{L}^\phi$. In that classical situation, to have uniqueness, one should require the so-called *finite kinetic energy condition* below (see [7], Th. 5.38, p. 188)

$$\text{(D.iv)} \qquad \mathbb{Q}\left(E_t(\phi) < +\infty,\ \forall t < \zeta\right) = 1.$$

Notice that (D.iv) is equivalent to $\mathbb{Q}(\tau_\infty = \zeta) = 1$ where $\tau_\infty := \sup_{n\ge1}\tau_n$.

The finite kinetic energy condition (D.iv) is indispensable for the martingale uniqueness in the actual setting as well. Indeed, if there were one probability $\mathbb{Q}$ satisfying (D.i) → (D.iii), except for (D.iv), then

$$\mathbb{Q}' := \mathbb{Q} \quad \text{on } [\tau_\infty = \zeta]; \quad \mathbb{Q}'(X_t = \Delta, \forall t > \tau_\infty | \tau_\infty < \zeta) = 1$$

satisfies still (D.i) $\rightarrow$ (D.iii).

The careful reader might be troubled with the above result by the point that (D.iv) is not used. Indeed we have spent a lot of time in coming to realize that (D.iv) is a consequence of (D.ii) and our strong condition (2.6).

To have the uniqueness of $\mathcal{L}^\phi$ without the local absolute continuity condition (2.6), we introduce:

(H_ν) **Local uniqueness for** $(\mathcal{L}, \mathcal{D}, \nu)$: *for any* $(\mathcal{F}_t)$ *stopping time* $\tau \le \zeta$, *if* $\mathbb{Q}$ *is a solution to the martingale problem associated with* $(\mathcal{L}, \mathcal{D}, \nu)$ *until* τ, *i.e.,* $\mathbb{Q}(X_{\tau\wedge\cdot} \in \cdot)$ *satisfies (D.i), (D.ii), (D.iii) with* $\phi = 1$), *then* $\mathbb{Q} = \mathbb{P}_\nu$ *over* $\mathcal{F}_\tau$.

This definition is slightly different from the classical one given in [7] (Def. 2.35, p. 146), where only $(\mathcal{F}_t^0)$-stopping times are used. The main reason that we take $(\mathcal{F}_t)$-stopping times in the assumption above is the fact that the (τ_n), defined in (2.2), are $(\mathcal{F}_t)$-stopping time, but we dont know whether they are so w.r.t. $(\mathcal{F}_t^0)$.

Theorem 2.3. *Assume* (H_ν) *for some given initial measure* $\nu \ll \mu_\phi$. *If* $\mathbb{Q}$ *is a solution to the martingale problem associated with* $(\mathcal{L}^\phi, \mathcal{D}, \nu)$, *such that*

$$\mathbb{Q}_t(A) := \mathbb{Q}(X_t \in A) \le C(t)\mu_\phi(A), \quad \forall t \ge 0, \quad \forall A \in \mathcal{B} \tag{2.8}$$

where $C(\cdot) \ge 0$ *is some finite nondecreasing function on* $\mathbb{R}^+$, *then for each* $T > 0$, $\mathbb{Q}|_{\mathcal{F}_T^0} = \mathbb{Q}_\nu^\phi|_{\mathcal{F}_T^0}$ *given by (2.7).*

Notice that condition (2.8) is much stronger than the finite kinetic energy condition (D.iv).

We now apply the two previous results to the L^p-Markov uniqueness defined as follows:

Definition 2.4. Let $1 \le p < +\infty$ and assume that $\mathcal{L}^\phi(\mathcal{D}) \subset L^p(E, \mu_\phi)$. $(\mathcal{L}^\phi, \mathcal{D})$ is called $L^p(\mu_\phi)$-*Markov unique*, if there is only one sub-Markov C_0-semigroup on $L^p(E, \mu_\phi)$ such that its generator in $L^p(E, \mu_\phi)$ is an extension of $(\mathcal{L}^\phi, \mathcal{D})$.

For a sub-Markov C_0-semigroup (Q_t) on $L^p(E, \mu_\phi)$ whose generator is an extension of $(\mathcal{L}^\phi, \mathcal{D})$, it is legitimate to think that the Markov process law $\mathbb{Q}$ with (Q_t) as transition semigroup and μ_ϕ as initial measure solves the martingale problem $(\mathcal{L}^\phi, \mathcal{D}, \mu_\phi)$. Consequently the following result is natural (but not so easy).

Theorem 2.5. *Let* $p \in [1, +\infty)$ *and assume that* $\mathcal{L}^\phi(\mathcal{D}) \subset L^p(E, \mu_\phi)$.

(a) If the martingale problem $(\mathcal{L}^\phi, \mathcal{D}, \mu_\phi)$ *has a unique solution, then* $(\mathcal{L}^\phi, \mathcal{D})$ *is* $L^p(\mu_\phi)$*-Markov unique;*

(b) If (2.6) holds for every solution $\mathbb{Q}$ *of the martingale problem* $(\mathcal{L}^\phi, \mathcal{D}, \mu_\phi)$, *then* $(\mathcal{L}^\phi, \mathcal{D})$ *is* $L^p(\mu_\phi)$*-Markov unique;*

(c) If $p = 1$ *and* (H_{μ_ϕ}) *holds, then* $(\mathcal{L}^\phi, \mathcal{D})$ *is* $L^1(\mu_\phi)$*-Markov unique.*

In all three cases $(\mathcal{L}^\phi, \mathcal{D})$ *is e.s.m.a.*

Remarks (2.iii). Because a strongly continuous **symmetric** sub-Markov semigroup on $L^2(E, \mu_\phi)$ is a sub-Markov C_0-semigroup on $L^p(E, \mu_\phi)$ for all $1 \le p < +\infty$, the L^p-Markov uniqueness for any $p \in [1, +\infty)$ is stronger than the usual e.m.s.a. The last claim above is then obvious.

Moreover for $1 \le p < q < +\infty$, since a sub-Markov C_0-semigroup on $L^p(E, \mu_\phi)$ is so on $L^q(E, \mu_\phi)$ (by the Riesz–Torin interpolation theorem), then the L^q-Markov uniqueness is stronger than L^p-Markov uniqueness. In other words, the L^1-Markov uniqueness in Theorem 2.5.(c) is the weakest. It is very interesting to check the L^2-Markov uniqueness in Theorem 2.5.(c).

Remarks (2.iv). The result above for $p > 2$ is not very interesting because the condition $\mathcal{L}^\phi(\mathcal{D}) \subset L^p(E, \mu_\phi)$ becomes very restrictive for the wave function, particularly for its behavior near the node set. Recall that condition $\mathcal{L}^\phi(\mathcal{D}) \subset L^p(E, \mu_\phi)$ is satisfied for $1 \le p \le 2$.

Remarks (2.v). When $(E, \mathcal{L}, \mathcal{D}) = (\mathbb{R}^d, \Delta/2, C_0^\infty(\mathbb{R}^d))$, condition (2.6) is satisfied automatically; thus for any $\phi \in H^1(\mathbb{R}^d)$, $(\mathcal{L}^\phi, \mathcal{D})$ is L^2-Markov unique. This strenthens the previous result of Röckner and Zhang [11], [12].

For the infinite dimensional models treated by Song [13] and Röckner and Zhang [11], [12], we can still show that (2.6) is satisfied for any martingale solution. Then the L^2-Markov uniqueness in Theorem 2.5 above holds, strenthening their results.

3 Proof of Theorem 2.2

We require two useful observations below. The first is well known:

Lemma 3.1. *For any $u \in \mathcal{D}$, let*

$$M_t(u) := u(X_t) - u(X_0) - \int_0^t \mathcal{L}u(X_s)ds, \ t \ge 0 \tag{3.1}$$

be the assiciated $\mathbb{P}_\nu$-martingale. Then for all $u, v \in \mathbf{D}(\mathcal{E})$ and $\nu \ll \mu$, the predictable quadratic covariational process between $M(u)$ and $M(v)$ (under $\mathbb{P}_\nu$) is given by

$$\langle M(u), M(v)\rangle_t = 2\int_0^t \Gamma(u, v)(X_s)ds, \ \forall t \ge 0, \ \mathbb{P}_\nu\text{-a.s.}$$

In particular, $\langle L^\phi\rangle_t = 2E_t(\phi)$, $\forall t < \sigma^\phi$, $\mathbb{P}_\nu$-a.s.

Lemma 3.2. *For the fixed Borel μ-versions ϕ and $\Gamma(\phi) \ge 0$, we have*

(a) $t \mapsto E_t(\phi)$ is continuous on $[0, \tau_\infty)$ where $\tau_\infty = \sup_{n\ge 1} \tau_n$, left continuous on $[0, \zeta]$, and

$$E_{\tau_n}(\phi) \le n.$$

(b) $\tau_n, n \geq 1$ defined by (2.2) are $(\mathcal{F}_t)$-stopping times;

(c) If $\mathbb{Q}$ satisfies (D.ii), then (2.4) and (2.5) hold $\mathbb{Q}$-a.s

Proof. **a)** Since $\tau_\infty = \inf\{t \in [0, \zeta) | E_t(\phi) = +\infty\}$, then by dominated convergence $E_\cdot(\phi)$ is continuous on $[0, \tau_\infty)$. By monotone convergence $E_\cdot(\phi)$ is left continuous on $[0, \zeta]$. For the last claim, it holds trivially on $[\tau_n = 0]$. Now on $[\tau_n > 0]$, if $t < \tau_n$, then $E_t(\phi) \leq n$. By monotone convergence,

$$E_{\tau_n}(\phi) = \int_0^{\tau_n} \frac{\Gamma(\phi)}{\phi^2}(X_s)ds = \lim_{t \uparrow \tau_n} \int_0^t \frac{\Gamma(\phi)}{\phi^2}(X_s)ds = \lim_{t \uparrow \tau_n} E_t(\phi) \leq n.$$

b) Fix $t \geq 0$. Obviously $[E_t(\phi) \leq n] \subset [t \leq \tau_n]$ by definition (2.2) of τ_n. But $[t \leq \tau_n] \subset [E_t(\phi) \leq n]$ by part a). Then $[\tau_n \geq t] \in \mathcal{F}_t^0$ for all $t \geq 0$. Consequently $[\tau_n > t] = \bigcap_{m \geq 1} [\tau_n \geq t + 1/m] \in \bigcap_{m \geq 1} \mathcal{F}^0_{t+1/m} = \mathcal{F}_t$, the desired result.

c) For all $t \leq \tau_n$, $E_t(\phi) \leq n$. Thus by Cauchy–Schwartz, we have for all $u \in \mathcal{D}$ and for all $t \leq \tau_n$,

$$\left(\int_0^t \left|\frac{\Gamma(\phi, u)}{\phi}(X_s)\right| ds\right)^2 \leq \int_0^t \frac{\Gamma(\phi)}{\phi^2}(X_s)ds \cdot \|\Gamma(u)\|_\infty t \leq nt\|\Gamma(u)\|_\infty$$

where (2.4) follows. Now the equality (2.5) follows from (2.4) by Fubini's theorem. ∎

Warning. it is possible that $E_{\tau_n+}(\phi) = +\infty$ on $[\tau_n < \zeta]$!

We turn now to the proof of Theorem 2.2.

We begin by a general remark: if ν, ν' are equivalent, for any solution $\mathbb{Q}$ to the martingale problem of $(\mathcal{L}, \mathcal{D}, \nu)$, $(d\nu'/d\nu)(X_0) \cdot \mathbb{Q}$ is a solution to $(\mathcal{L}, \mathcal{D}, \nu')$. Hence we can assume without loss of generality that $d\nu/d\mu \in L^\infty(\mu)$. The proof is separated into four steps.

Step 1. Fix $T > 0, n \geq 1$. In this step by using an elementary (and standard) argument, we translate the absolute continuity condition (2.6) into an explicit (Girsanov) formula (3.4) below for the density process

$$D_t := \frac{d\mathbb{Q}}{d\mathbb{P}_\nu}|_{\mathcal{F}_{t \wedge \tau_n \wedge T}}, t \geq 0. \tag{3.2}$$

Remark that $D_0 = 1, \mathbb{P}_\nu$-a.e. By our continuous path assumption on the free process $(\mathbb{P}_x)_{x \in E}$, the $\mathbb{P}_\nu$-martingale $(D_t)_{t \geq 0}$ can be chosen to be *continuous*, that will be assumed below.

Let $\sigma_\epsilon = \inf\{t \geq 0; D_t \leq \epsilon\}, \sigma = \sup_{\epsilon > 0} \sigma_\epsilon$. For $t \leq \sigma_\epsilon$, set

$$L_t = \int_0^t \frac{1}{D_s} dD_s. \tag{3.3}$$

By Ito's formula, for all $t \le \sigma_\epsilon$,

$$\log D_t = \int_0^t \frac{1}{D_s} dD_s - \frac{1}{2}\int_0^t \frac{1}{(D_s)^2} d\langle D\rangle_s = L_t - \frac{1}{2}\langle L\rangle_t$$

we get hence

$$\frac{d\mathbb{Q}}{d\mathbb{P}_\nu}|_{\mathcal{F}^\nu_{t\wedge\tau_n\wedge T\wedge\sigma_\epsilon}} = D_{t\wedge\sigma_\epsilon} = \exp\left(L_t^{\sigma_\epsilon} - \frac{1}{2}\langle L^{\sigma_\epsilon}\rangle_t\right). \tag{3.4}$$

where $L_t^S = L_{t\wedge S}$ for any stopping time S.

Step 2. Introduce now

$$\sigma_m^\phi := \inf\left\{t \ge 0; \tilde{\phi}(X_t) \le \frac{1}{m}\right\} \tag{3.5}$$

and write $S = T \wedge \tau_n \wedge \sigma_\epsilon \wedge \sigma_m^\phi$. Remark that $\mathbb{P}_\nu$, $\mathbb{Q}$ are equivalent on $\mathcal{F}_S^\nu$ by (3.4) (S is stopping time w.r.t. $(\mathcal{F}_t^\nu)$, but not w.r.t. $(\mathcal{F}_t)$!).

For every $u \in \mathcal{D}$, since $M(u)$ given by (3.1) is a $\mathbb{P}_\nu$-martingale, then

$$M_{\cdot\wedge S}(u) - \langle M^S(u), L^S\rangle.$$

is a $\mathbb{Q}$-local martingale by Girsanov formula [JS, p. 155, Th. 3.11]. On the other hand, by (D.iii),

$$M_{t\wedge S}(u) - 2\int_0^{t\wedge S} \frac{\Gamma(\phi, u)}{\phi}(X_s)ds = M^\phi_{t\wedge S}(u)$$

is a $\mathbb{Q}$-local martingale too. Consequently $\mathbb{Q}$-a.s. for all $t \ge 0$

$$\langle L^S, M(u)\rangle_t = 2\int_0^{t\wedge S} \frac{\Gamma(\phi, u)}{\phi}(X_s)ds. \tag{3.6}$$

By the equivalence of $\mathbb{P}_\nu$ and $\mathbb{Q}$ on $\mathcal{F}_S$, (3.6) holds $\mathbb{P}_\nu$-a.s.

Return now to the martingale $L_t^{\phi,S} = \int_0^{t\wedge S}(1/\tilde{\phi})(X_s)dM_s(\phi)$ given in (1.10). By Lemma 3.1 and (2.5), we have $\mathbb{P}_\nu$-a.s.,

$$\langle L^{\phi,S}, M(u)\rangle_t = \int_0^{t\wedge S} \frac{1}{\tilde{\phi}(X_s)} d\langle M(\phi), M(u)\rangle_s = 2\int_0^{t\wedge S} \frac{\Gamma(\phi, u)}{\phi}(X_s)ds.$$

Combining this equality with (3.6) we get

$$\langle L^S - L^{\phi,S}, M(u)\rangle = 0, \ \mathbb{P}_\nu\text{-a.s.} \quad \forall u \in \mathcal{D}. \tag{3.7}$$

Step 3. In this step we shall prove that (3.7) implies the key equality $L^S = L^{\phi,S}$ ($\mathbb{P}_\nu$-a.s.).

Notice that $u \mapsto M(u)$ is continuous from $\mathbf{D}(\mathcal{E})$ to the space $\mathcal{M}^2(\mathbb{P}_\nu)$ of the $L^2(\mathbb{P}_\nu)$-martingales (since $d\nu/d\mu \in L^\infty$). Since $\mathcal{D}$ is a core for $(\mathcal{E}, \mathbf{D}(\mathcal{E}))$, by continuous extension the above relation (3.7) still holds for all $u \in G_1(b\mathcal{B} \cap L^2(\mu)) \subset \mathbf{D}(\mathcal{E})$, where $b\mathcal{B}$ is the space of all real Borel measurable bounded functions on E, and

$$G_1 f(x) := \mathbb{E}^x \int_0^{+\infty} e^{-t} f(X_t) dt$$

is the resolvent associated with the free process $(\mathbb{P}_x)_{x \in E}$.

Next let $(E_k \in \mathcal{B})_{k \geq 1}$ be an increasing sequence such that $\bigcup_k E_k = E$ and $\mu(E_k) < +\infty$. For each $f \in b\mathcal{B}$, $f_k := f 1_{E_k} \in b\mathcal{B} \cap L^2(\mu)$. For each $t > 0$ fixed, we have

$$\begin{aligned} M_t(G_1 f_k) &= G_1 f_k(X_t) - G_1 f_k(X_0) - \int_0^t (G_1 f_k - f_k)(X_s) ds \\ &\longrightarrow G_1 f(X_t) - G_1 f(X_0) - \int_0^t (G_1 f - f)(X_s) ds = M_t(G_1 f) \end{aligned}$$

in probability $\mathbb{P}_\nu$ and then in $L^2(\mathbb{P}_\nu)$, by the dominated convergence and the fact that $\|M_t(G_1 f_k)\|_\infty \leq 2(1+t)\|f\|_\infty$. Consequently (3.7) holds for all $u \in G_1(b\mathcal{B})$.

Now by the classical Kunita–Watanabe theorem [5] (Chap. XV, p. 243, Th. 19]), the *stable* subspace generated by $\{M(G_1 f);\ f \in b\mathcal{B}\}$ contains all $\mathbb{P}_\nu$-local martingales issued of zero. Thus (3.7) for all $u \in G_1(b\mathcal{B})$ implies

$$L^S - L^{\phi,S} = 0, \quad \mathbb{P}_\nu\text{-a.s.} \tag{3.8}$$

Consequently we can rewrite (3.4) as

$$\frac{d\mathbb{Q}}{d\mathbb{P}_\nu}|_{\mathcal{F}_S} = \exp\left(L_S^\phi - \frac{1}{2}\langle L^\phi\rangle_S\right) = \exp\left(L_S^\phi - E_S(\phi)\right). \tag{3.9}$$

Step 4. Having the key fact (3.9), the remained part of our proof is routine (but uses Theorem A in a crucial way).

Recall that $S = T \wedge \tau_n \wedge \sigma_\epsilon \wedge \sigma_m^\phi$. When $\epsilon \downarrow 0$, $\sigma_\epsilon \uparrow \sigma$ in (3.9). Since

$$E_{T \wedge \tau_n \wedge \sigma_m^\phi \wedge \sigma_\epsilon}(\phi) \leq E_{\tau_n \wedge \sigma_m^\phi \wedge \sigma}(\phi) \leq n$$

for all $\epsilon > 0$ (by Lemma 3.2), then (3.9) continues to hold for $S_2 = T \wedge \tau_n \wedge \sigma \wedge \sigma_m^\phi$ (instead of S), by Novikov's criterion. Thus $\mathbb{Q} \sim \mathbb{P}_\nu$ on $\mathcal{F}_{S_2}^\nu$.

Now by definition of σ, it is well known that $\mathbb{Q}(\sigma = +\infty) = 1$. Then by the equivalence shown previously and the fact that $[\sigma < T \wedge \tau_n \wedge \sigma_m^\phi] \in \mathcal{F}_{S_2}^\nu$, we have

$$\mathbb{P}_\nu\left(\sigma < T \wedge \tau_n \wedge \sigma_m^\phi\right) = 0, \text{ i.e., } S_2 = T \wedge \tau_n \wedge \sigma_m^\phi,\ \mathbb{P}_\nu\text{-a.s.} \tag{3.10}$$

Hence by (3.8) for S_2, we get

$$\mathbb{Q} = e(L^\phi)_{T\wedge\tau_n\wedge\sigma_m^\phi} d\mathbb{P}_\nu = \mathbb{Q}_\nu^\phi \quad \text{over } \mathcal{F}^\nu_{T\wedge\tau_n\wedge\sigma_m^\phi}. \tag{3.11}$$

By the *a priori* node probability estimation (1.13) and the trivial fact that $\mathbb{Q}^\phi_{\mu_\phi}(\tau_\infty = +\infty) = 1$,

$$\mathbb{Q}^\phi_\nu\left(\sigma^\phi \wedge \tau_\infty = \sigma^\phi = +\infty\right) = 1$$

then by (3.11),

$$\mathbb{Q}\left(T < \tau_n \wedge \sigma_m^\phi\right) = \mathbb{Q}^\phi_\nu\left(T < \tau_n \wedge \sigma_m^\phi\right) \longrightarrow 1 \tag{3.12}$$

as $n, m \to \infty$. Now for every $A \in \mathcal{F}_T$, we get by (3.12)

$$\begin{aligned}
\mathbb{Q}(A) &= \lim_{m,n\to\infty} \mathbb{Q}\left(A \bigcap [\tau_n \wedge \sigma_m^\phi > T]\right) \\
&= \lim_{m,n\to\infty} \mathbb{Q}^\phi_\nu\left(A \bigcap [\tau_n \wedge \sigma_m^\phi > T]\right) \\
&= \mathbb{Q}^\phi_\nu(A),
\end{aligned}$$

the desired result. ∎

Remarks (3.i). The absolute continuity condition (2.6) is in some sense indispensable for the martingale uniqueness in Theorem 2.1. In fact return to the non-perturbed case where μ is a probability and $\phi = 1$. In Step 3 we have proved

Proposition 3.3. *If $\mathcal{D}$ is a core for $\mathbf{D}(\mathcal{E})$, then the stable subspace generated by $\{M(u);\, u \in \mathcal{D}\}$ contains all $\mathbb{P}_\nu$-local martingales issued of zero.*

The last conclusion above is equivalent to: $\mathbb{P}_\nu$ is an extremal element in the convex set of the solutions to the martingale problem associated with $(\mathcal{L}, \mathcal{D}, \nu)$, by [5] (Chap. VIII, Th. 57). So only for those martingale solutions $\mathbb{Q}$ of $(\mathcal{L}, \mathcal{D}, \nu)$ which are moreover absolutely continuous w.r.t. $\mathbb{P}_\nu$, we can conclude that $\mathbb{Q} = \mathbb{P}_\nu$.

As the reader see clearly, Kunita-Watanabe's theorem in Step 3 above plays a crucial role.

4 Proof of Theorem 2.3

The main idea consists of constructing a $\mathbb{Q}$-local martingale M^{τ_n} which seems like

$$M^\phi_{t\wedge\tau_n}(\log\phi) = \int_0^{t\wedge\tau_n} \phi^{-1}(X_s)\, dM^\phi(\phi)$$

to show next that

$$\exp\left(-M_t^{\tau_n} - E_{t\wedge\tau_n}(\phi)\right) \cdot \mathbb{Q}|_{\mathcal{F}_{t\wedge\tau_n}}$$

gives a solution to the martingale problem associated with $(\mathcal{L}, \mathcal{D}, \nu)$ until τ_n. So (H_ν) implies the equivalence of $\mathbb{Q}$ and $\mathbb{P}_\nu$ on $\mathcal{F}_{t\wedge\tau_n}$. It remains to apply Theorem 2.1.

We shall realize this idea in two steps.

Step 1. The purpose of this step is to construct a $\mathbb{Q}$-local martingale possessing all properties of M^ϕ $(\log\phi)$. This is quite delicate, because *a priori* M_t^ϕ $(\log\phi)$ is well defined only $\mathbb{P}_\nu$-a.s. until $\tau_\infty \wedge \sigma^\phi$ and we do not know whether $\mathbb{Q} \ll \mathbb{P}_\nu$ even locally, and σ^ϕ is not $\mathbb{Q}$-well defined.

Our construction below consists of approximating $M^\phi(\phi)$ by $M^\phi(u)$ where $u \in \mathcal{D}$. First, for each $u \in \mathcal{D}$, $u^2 \in \mathcal{D}$, we have μ_ϕ-a.s.,

$$\begin{aligned}\mathcal{L}^\phi(u^2) &= \mathcal{L}(u^2) + 2\frac{\Gamma(\phi, u^2)}{\phi}\\ &= 2u\mathcal{L}u + 2\Gamma(u,u) + 4u\frac{\Gamma(\phi,u)}{\phi}\\ &= 2u\mathcal{L}^\phi u + 2\Gamma(u,u)\end{aligned}$$

Then

$$M^\phi_{\tau_n\wedge\cdot}(u^2) = u^2(X_{\tau_n\wedge\cdot}) - u^2(X_0) - \int_0^{\tau_n\wedge\cdot} (2u\mathcal{L}^\phi u + 2\Gamma(u,u))(X_s)ds$$

is $\mathbb{Q}$-local martingale. On the other hand by Ito's formula, we have

$$\begin{aligned}u^2(X_{\tau_n\wedge\cdot}) - u^2(X_0) &= 2\int_0^{\tau_n\wedge\cdot} u(X_s)\big(dM_s^\phi(u)\\ &\quad + \mathcal{L}^\phi u(X_s)ds\big) + \langle M^\phi(u)\rangle_{\tau_n\wedge\cdot}\,.\end{aligned}$$

By comparing the previous two equalities we get $\mathbb{Q}$-a.s.,

$$\langle M^\phi(u)\rangle_t = 2\int_0^t \Gamma(u,u)(X_s)ds, \quad \forall t \le \tau_n. \tag{4.1}$$

By polarization, we have for all $u, v \in \mathcal{D}$, $\mathbb{Q} - a.s.$,

$$\langle M^\phi(u), M^\phi(v)\rangle_t = 2\int_0^t \Gamma(u,v)(X_s)ds, \quad \forall t \le \tau_n. \tag{4.2}$$

Now fix a $(\mathcal{F}_t^{\mathbb{Q}})$-predictable $dt \otimes \mathbb{Q}$-version $Y.$ of $\phi(X.)$. We choose $(u_k) \subset \mathcal{D}$ such that $\mathcal{E}_1(u_k - \phi) \to 0$.

Notice that by (D.ii) and our condition (2.8),

$$\begin{aligned}\mathbb{E}^{\mathbb{Q}}\int_0^{\tau_n\wedge t} \frac{1}{Y_s^2} d\langle M^\phi(u_k)\rangle_s &= 2\mathbb{E}^{\mathbb{Q}}\int_0^{\tau_n\wedge t} \frac{\Gamma(u_k,u_k)(X_s)}{Y_s^2}ds\\ &= 2\mathbb{E}^{\mathbb{Q}}\int_0^{\tau_n\wedge t} \frac{\Gamma(u_k,u_k)}{\phi^2}(X_s)ds\\ &\le 2C(t)\int_0^t\int_E \Gamma(u_k,u_k)\frac{1}{\phi^2}\cdot\phi^2 d\mu ds\\ &= 2C(t)t\mathcal{E}(u_k,u_k) < +\infty.\end{aligned}$$

We can then define the stochastic integral below under $\mathbb{Q}$,

$$M_t^k := \int_0^{t\wedge\tau_n} \frac{1}{Y_s} dM_s^\phi(u_k).$$

It is a $L^2(\mathbb{Q})$-martingale (w.r.t. $(\mathcal{F}_t^{\mathbb{Q}})$). By (4.1) and (4.2), we have

$$\langle M^k\rangle_t = 2\int_0^{t\wedge\tau_n} \frac{\Gamma(u_k)}{\phi^2}(X_s)ds \tag{4.3}$$

$$\langle M^k - M^l\rangle_t = 2\int_0^{t\wedge\tau_n} \frac{\Gamma(u_k - u_l)}{\phi^2}(X_s)ds \tag{4.4}$$

$$\langle M^k, M^\phi(u)\rangle_t = 2\int_0^{t\wedge\tau_n} \frac{\Gamma(u_k, u)}{\phi}(X_s)ds \tag{4.5}$$

all $\mathbb{Q}$-a.s. Still by our condition (2.8), we get from (4.4)

$$\mathbb{E}^{\mathbb{Q}}\langle M^k - M^l\rangle_t \le 2C(t)t\mathcal{E}(u_k - u_l) \longrightarrow 0$$

as $k, l \to \infty$. Then there is a $L^2(\mathbb{Q})$-continuous martingale $M_\cdot^{\tau_n}$ such that

$$\mathbb{E}^{\mathbb{Q}}\langle M^k - M^{\tau_n}\rangle_t \longrightarrow 0, \quad \forall t \ge 0. \tag{4.6}$$

By (4.3) and that fact that $\int \Gamma(u_k - \phi)d\mu \to 0$, the convergence (4.6) above implies as before that

$$\langle M^{\tau_n}, M^{\tau_n}\rangle_t = 2\int_0^{t\wedge\tau_n} \frac{\Gamma(\phi)}{\phi^2}(X_s)ds = 2E_{t\wedge\tau_n}(\phi), \ \mathbb{Q}\text{-a.s.} \tag{4.7}$$

We now look at the limit of (4.5) when $k \to \infty$. To this end, notice that

$$\begin{aligned} &\mathbb{E}^{\mathbb{Q}} \int_0^{t\wedge\tau_n} \left(\frac{\Gamma(u_k - \phi, u)}{\phi}\right)^2 (X_s)ds \\ &\quad \le \mathbb{E}^{\mathbb{Q}} \int_0^{t\wedge\tau_n} \frac{\Gamma(u_k - \phi)}{\phi^2}(X_s)ds \cdot \|\Gamma(u)\|_\infty \\ &\quad \le C(t)t \cdot \mathcal{E}(u_k - \phi) \cdot \|\Gamma(u)\|_\infty \longrightarrow 0. \end{aligned}$$

By this estimation and (4.6), the relation (4.5) becomes as $k \to \infty$

$$\langle M^{\tau_n}, M^\phi(u)\rangle_t = 2\int_0^{t\wedge\tau_n} \frac{\Gamma(\phi, u)}{\phi}(X_s)ds \tag{4.8}$$

$\mathbb{Q}$-a.s. The so constructed M^{τ_n} will be our ($\mathbb{Q}$)-local martingale promised in the outline of the proof.

Step 2. Now on $\mathcal{F}_{T\wedge\tau_n}$ ($T \in \mathbb{R}^+$), define a new probability

$$\mathbb{P}^n = \exp\left(-M_T^{\tau_n} - \frac{1}{2}\langle M^{\tau_n}\rangle_T\right) \cdot \mathbb{Q}. \tag{4.9}$$

Since $\langle M^{\tau_n}\rangle_T = 2E_{T\wedge\tau_n}(\phi) \le 2n$ by (4.7), then by Novikov's criterion

$$\left(\exp\left(-M_T^{\tau_n} - \frac{1}{2}\langle M^{\tau_n}\rangle_T\right)\right)_T$$

is actually a $\mathbb{Q}$-martingale. Hence (4.9) well defines a probability on $\mathcal{F}_{T\wedge\tau_n}$.

By Girsanov's formula ([JS, pp. 155–156, Th. 3.11], for each $u \in \mathcal{D}$,

$$Z := M^{\phi}_{\cdot\wedge\tau_n}(u) + \langle M^{\tau_n}, M^{\phi}(u)\rangle.$$

is a $\mathbb{P}^n$-local martingale. But by (4.8),

$$\begin{aligned} Z_t &= u(X_{t\wedge\tau_n}) - u(X_0) - \int_0^{t\wedge\tau_n} \mathcal{L}^{\phi}u(X_s)ds + 2\int_0^{t\wedge\tau_n} \frac{\Gamma(\phi,u)}{\phi}(X_s)ds \\ &= u(X_{t\wedge\tau_n}) - u(X_0) - \int_0^{t\wedge\tau_n} \mathcal{L}u(X_s)ds. \end{aligned}$$

In other words, $\mathbb{P}^n$ is a solution to the martingale problem associated to $(\mathcal{L}, \mathcal{D}, \nu)$ over $[0, \tau_n]$.

By our local uniqueness assumption (H_ν), $\mathbb{P}^n = \mathbb{P}_\nu$ over $\mathcal{F}_{T\wedge\tau_n}$. Namely $\mathbb{Q}$ is equivalent to $\mathbb{P}_\nu$ over $\mathcal{F}_{T\wedge\tau_n}$. Then $\mathbb{Q}$ satisfies the condition (2.6). It remains to apply Theorem 2.1. ■

5 Proof of Theorem 2.5

a) Let (Q_t) be a sub-Markov C_0-semigroup on $L^p(E, \mu_\phi)$ such that its generator $(A, \mathbf{D}(A)) \supset (\mathcal{L}^{\phi}, \mathcal{D})$. We extend it to $L^p(E_\Delta, \mu_\phi^\Delta := \mu_\phi + \delta_\Delta)$ in the usual way:

$$Q_t^\Delta f(x) = Q_t(f1_E)(x) + (1 - Q_t 1_E(x))f(\Delta), \quad \forall f \in b\mathcal{B}_\Delta.$$

$Q_t^\Delta f$ is μ_ϕ^Δ–well defined, $\mu_\phi^\Delta Q_t^\Delta \ll \mu_\phi^\Delta$ and (Q_t^Δ) is a sub-Markov C_0-semigroup on $L^p(E_\Delta, \mu_\phi^\Delta)$.

Hence on the product space $(E_\Delta^{\mathbb{R}^+}, \mathcal{B}_\Delta^{\mathbb{R}^+})$, there is a unique probability $\mathbb{Q}$ such that the coordinates process $(Y_t)_{t\ge0}$ is a Markov process with initial measure $\mu_\phi \ll \mu_\phi^\Delta$ and transition semigroup (Q_t^Δ).

The reader see clearly what we should do below for proving that $\mathbb{Q}$ is a solution to the martingale problem of $(\mathcal{L}^\phi, \mathcal{D}, \nu)$ given in Definition 2.1: the only difficulty is to show that $\mathbb{Q}$ is supported by $\Omega = \Omega^{\mathcal{D}}$ given at the beginning of §2. We will prove this by four points.

In the following any $[-\infty, +\infty]$-valued function f on E will be regarded as a function on E_Δ with $f(\Delta) = 0$.

1) First we claim that (Y_t) admits one measurable version $(\bar{Y}_t)_{t\ge0}$, i.e., $(t,\omega) \mapsto \bar{Y}_t(\omega)$ is $\mathcal{B}(\mathbb{R}^+) \times \mathcal{F}/\mathcal{B}_\Delta$-measurable, where $\mathcal{F} = (\mathcal{B}_\Delta^{\mathbb{R}^+})^{\mathbb{Q}}$, the completion w.r.t. $\mathbb{Q}$. In fact let $\eta : E_\Delta \to [0, 1]$ be an injection bimeasurable (it exists since E is Polish). Since

(i) $s \mapsto \mathbb{E}^{\mathbb{Q}}\eta(Y_s)^2 = \int_{E_\Delta} Q_s^\Delta(\eta^2)d\mu_\phi$ is continuous by the strong continuity of (Q_s^Δ) on $L^p(E_\Delta, \mu_\phi^\Delta)$;

(ii) $s \mapsto \mathbb{E}^{\mathbb{Q}}\eta(Y_s)f_1(Y_{t_1})\cdots f_k(Y_{t_k})$ is continuous, where $f_k \in b\mathcal{B}_\Delta$, $t_1 < \cdots < t_k$ are arbitrary (that follows easily from the Markov property and the strong continuity of (Q_t^Δ) in L^p);

hence $s \mapsto \eta(Y_s) \in L^2(\mathbb{Q})$ is continuous. By Doob's theorem (see [5], Chap. IV, Th. 30, p. 158), the process $(\eta(Y_s))$ admits one *measurable* version (in the sense above). Consequently (Y_s) does too.

2) On $(E_\Delta^{\mathbb{R}^+}, \mathcal{F}, \mathbb{Q})$, define for each $u \in \mathcal{D}$,

$$\bar{M}_t^\phi(u) := u(\bar{Y}_t) - u(\bar{Y}_0) - \int_0^t \left(\mathcal{L}u + 2\frac{\Gamma(\phi, u)}{\phi}\right)(\bar{Y}_s)ds \tag{5.1}$$

where the last integral is a well-defined continuous process because $\mathcal{L}^\phi u \in L^p(\mu_\phi)$. Since the generator $(A, \mathbf{D}(A))$ of (Q_t) is an extension of $(\mathcal{L}^\phi, \mathcal{D})$, then $(\bar{M}_t^\phi(u))_{t\geq 0}$ is a $L^p(\mathbb{Q})$-martingale, which admits a càdlàg version. Moreover for each integer $m \geq 2$, we have

$$\begin{aligned}
\mathcal{L}^\phi(u^m) &= \mathcal{L}(u^m) + 2\frac{\Gamma(\phi, u^m)}{\phi} \\
&= mu^{m-1}\mathcal{L}u + m(m-1)u^{m-2}\Gamma(u) + 2mu^{m-1}\frac{\Gamma(\phi, u)}{\phi} \\
&= mu^{m-1}\mathcal{L}^\phi u + m(m-1)u^{m-2}\Gamma(u).
\end{aligned}$$

By a result of Bakry–Emery [2] (Lemma 2), for each $u \in \mathcal{D}$, $\bar{M}^\phi(u)$ admits a continuous $\mathbb{Q}$-version $\tilde{M}^\phi(u)$ such that

$$\langle \tilde{M}^\phi(u)\rangle_t = 2\int_0^t \Gamma(u)(Y_s)ds, \quad \mathbb{Q}\text{-a.s.} \tag{5.2}$$

3) Now take a countable dense subset $\mathbb{D} \subset \mathbb{R}^+$ fixed and a countable subset $\mathcal{D}_0 := \{f_k | k \geq 1\} \subset \mathcal{D}$ that generates $\mathcal{D}$ as the algebra (by (C.i)). By the second point 2) above, $(u(\bar{Y}_t))_{t\geq 0}$ admits a continuous version. Therefore the set

$$W := \left\{\omega \in E_\Delta^{\mathbb{R}^+} \;\middle|\; \begin{array}{l} t \mapsto u(Y_t(\omega)) \text{ is uniformly continuous on} \\ \mathbb{D}\bigcap[0, N], \forall N > 0, u \in \mathcal{D}_0 \end{array}\right\}$$

is of $\mathbb{Q}$-probability one. For $\omega \in W$, the uniform continuity above holds for all $u \in \mathcal{D}$.

As the algebra $\mathcal{D}$ separates the points of E, the topology $\sigma_{\mathcal{D}}$ generated by $\mathcal{D}$ on E_Δ is Hausdorff. Remark that $\sigma_{\mathcal{D}} = \sigma_{\mathcal{D}_0}$ and $\sigma_{\mathcal{D}_0}$ is metrizable with metric

$$\rho_{\mathcal{D}_0}(x, y) := \sum_{k=1}^\infty \frac{1}{2^k}\frac{|f_k(x) - f_k(y)|}{1 + |f_k(x) - f_k(y)|}.$$

We shall complete E_Δ w.r.t. $\rho_{\mathcal{D}_0}$ to get $\bar{E}_\Delta$. For each $\omega \in W$, $t \mapsto Y_t(\omega))$ is uniformly continuous from $\mathbb{D}\bigcap[0, N]$ to $(\bar{E}_\Delta, \rho_{\mathcal{D}_0})$, for all $N > 0$.

For each $\omega \in W$, we define $\tilde{Y}_.(\omega) \in \bar{E}_\Delta$ as the continuous extension on $\mathbb{R}^+$ of $t(\in \mathbb{D}) \mapsto (Y_t(\omega))$ in $(\bar{E}_\Delta, \sigma_{\mathcal{D}_0})$. And put finally

$$Z_t(\omega) = \begin{cases} \tilde{Y}_t(\omega), & \forall t < S(\omega),\ \forall \omega \in W; \\ \Delta, & \forall t \geq S(\omega),\ \forall \omega \in W; \\ x_0, & \forall \omega \notin W \end{cases}$$

where $S := \inf\{t \geq 0; \tilde{Y}_t(\omega) \notin E\}$, and x_0 is some fixed point of E. It is obvious that $(u(Z_.))$ is a continuous $\mathbb{Q}$-version of $(u(Y_.)$ for any $u \in \mathcal{D}_0$, then for all $u \in \mathcal{D}$. Moreover $Z_{\mathbb{D}} = Y_{\mathbb{D}}$, $\mathbb{Q}$-a.s. Thus Z is a $\mathbb{Q}$-version of Y.

4) We modify now the $\bar{E}_\Delta$-valued continuous process Z,

$$Z_t^\Delta(\omega) := \begin{cases} Z_t(\omega), & \text{if } t < S(\omega) := \inf\{t \geq 0 | Z_t(\omega) \notin E\} \\ \Delta, & \text{if } t \geq S(\omega). \end{cases}$$

$\mathbb{Q}(Z^\Delta \in \cdot)$ is supported by $\Omega = \Omega^{\mathcal{D}}$, and it solves the martingale problem $(\mathcal{L}^\phi, \mathcal{D}, \nu)$, by (5.1). By our martingale uniqueness assumption, $\mathbb{Q}(Z^\Delta \in \cdot) = \mathbb{Q}^\phi_{\mu_\phi}(\cdot)$. By the conservativity of $\mathbb{Q}^\phi_{\mu_\phi}$, that implies $\mathbb{Q}(S = +\infty) = 1$. Consequently

$$\mathbb{Q}^\phi_{\mu_\phi}(\cdot) = \mathbb{Q}(Z^\Delta \in \cdot) = \mathbb{Q}(Z \in \cdot) = \mathbb{Q}(Y \in \cdot),$$

the desired result of (a).

b) Part (b) follows from part (a) and Theorem 2.2.

c) For part (c), notice that if (Q_t) is a sub-Markov C_0-semigroup on $L^1(\mu_\phi)$, then letting

$$C(t) := \sup_{s \in [0,t]} \|Q_s\|_{L^1(\mu_\phi) \to L^1(\mu_\phi)},$$

we have for any $A \in \mathcal{B}$ (using objects in the proof of (a) above),

$$\mathbb{Q}(Z_t^\Delta \in A) \leq \mathbb{Q}(Z_t \in A) = \mathbb{Q}(Y_t \in A) = \int Q_t(1_A) d\mu_\phi \leq C(t)\mu_\phi(A).$$

In other words, $\mathbb{Q}(Z^\Delta \in \cdot)$ satisfies moreover (2.8). Applying Theorem 2.3, we get $\mathbb{Q}^\phi_{\mu_\phi}(\cdot) = \mathbb{Q}(Z^\Delta \in \cdot)$, then part (c). ■

Acknowledgment. I am grateful to M. Röckner who drew my attention to the martingale uniqueness question treated in this paper during my visit at Bonn on 1993. This work was reported in the Colloque Stoch. Analysis and Math. Phys. held in Lisbon 1999, and in the *Seminar on stochastic processes* held in Beijing 1999. My thanks go to A.B. Cruzeiro, Z.M. Ma, J.A. Yan and J.C. Zembrini for their warm invitation and hospitality.

References

[1] S. Albeverio, J.G. Kondratiev, and M. Röckner, An approximate criterium of essential self-adjointness of Dirichlet operators, *Potential Analysis*, **1** (1992), 307–317.

[2] D. Bakry and M. Emery, Diffusions hypercontractives, *Séminaire de Proba. XIX*, Lecture Notes in Math. 1123, 1985, pp. 177–206.

[3] P. Cattiaux and Ch. Léonard, Minimization of Kullback information of diffusion processes, *Ann. Inst. H. Poincaré*, **30**(1) (1994), 83–132.

[4] P. Cattiaux and M. Fradon, Entropy, reversible diffusion processes and Markov uniqueness, *J. Funct. Anal.*, **138**(1) (1996), 243–272.

[5] C. Dellacherie and P.A. Meyer, *Probabilités et Potentiel*, Chaptitres I à IV, Hermann, 1976; Chaptitres V à VIII, Hermann, 1980; Chapitres XII à XVI, Hermann, 1987.

[6] M. Fukushima, Y. Oshima, and M. Takeda, *Dirichlet Spaces and Symmetric Markov Processes*, Walter de Gruyter, 1994.

[7] J. Jacod and A.N. Shiryaev, *Limit Theorems for Stochastical Processes*, Springer-Verlag, 1987.

[8] Z.M. Ma and M. Röckner, *Introduction to the Theory of (non symmetric) Dirichlet Forms*, Springer-Verlag, 1992.

[9] P.A. Meyer and W.A. Zheng, Construction du processus de Nelson reversible, *Sem. Probab. XIX*, Lecture Notes Math., No. 1123, 1984, pp. 12–26.

[10] E. Nelson, *Quantum Fluctuations*, Princeton Series in Physics, Princeton University Press, 1985.

[11] M. Röckner and T.S. Zhang, Uniqueness of generalized Schrödinger operators and applications, *J. Funct. Anal.*, **105** (1992), 187–231.

[12] M. Röckner and T.S. Zhang, Uniqueness of generalized Schrödinger operators and applications, II, *J. Funct. Anal.*, **119** (1994), 455–467.

[13] S.Q. Song, A study on Markovian maximality, change of probability and regularity, *Potential Anal.*, **3** (1994), 391–422.

[14] S.Q. Song, Markov uniqueness and essential self-adjointness of perturbed Ornstein-Uhlenbeck operators, *Osaka J. Math.*, **32** (1995), 823–832.

[15] M. Takeda, The maximum Markovian self-adjoint extension of generalized Schrödinger's operators, *J. Math. Soc. Japan*, **44** (1992), 113–130.

[16] M. Takeda, Two classes of extensions for generalized Schrödinger operators, *Potential Analysis*, **5** (1996), 1–13.

[17] N. Wielens, On the essential self-adjointness of generalized Schrödinger operators, *J. Funct. Anal.*, **61** (1985), 98–115.

[18] L.M. Wu, Feynman-Kac semigroups, ground state diffusions and large deviations, *J. Funct. Anal.*, **123**(1) (1994), 202–231.

[19] L.M. Wu, Uniqueness of Schödinger operateurs restricted to a domain, *J. Funct. Anal.*, **153** (1998), 276–319.

[20] L.M. Wu, Uniqueness of Nelson's diffusions, *Probab. Theory Rel. Fields*, **114** (1999), 549–585.

[21] L.M. Wu, Uniqueness of Nelson's diffusions (II): infinite dimensional setting and applications, *Potential Analysis*, **13** (2000), 269–301.

L.M. Wu
Laboratoire de Math. Appl.
Université Blaise Pascal, 63177 Aubière, France
email: wuliming@ucfma.univ-bpclermont.fr
and
Department of Mathematics, Wuhan University
430072 Hubei, China
email: lmwu@colmath.whu.edu.cn

Progress in Probability

Progress in Probability is designed for the publication of workshops, seminars and conference proceedings on all aspects of probability theory and stochastic processes, as well as their connections with and applications to other areas such as mathematical statistics and statistical physics. It acts as a companion series to *Probability and Its Applications,* a context for research level monographs and advanced graduate texts.

We encourage preparation of manuscripts in some form of TeX for delivery in camera-ready copy, which leads to rapid publications, or in electronic form for interfacing with laser printers or typesetters.

Proposals should be sent directly to the editors or to:
Birkhäuser Boston, 675 Massachusetts Avenue, Cambridge, MA 02139, U.S.A.

12 Çinlar/Chung/Getoor. Seminar on Stochastic Processes, 1985
13 Çinlar/Chung/Getoor/Glover. Seminar on Stochastic Processes, 1986
14 Devroye. A Course in Density Estimation
15 Çinlar/Chung/Getoor/Glover. Seminar on Stochastic Processes, 1987
16 Kiper. Random Perturbations of Dynamical Systems
17 Çinlar/Chung/Getoor/Glover. Seminar on Stochastic Processes, 1988
18 Çinlar/Chung/Getoor/Fitzsimmons/Williams. Seminar on Stochastic Processes, 1989
19 Alexander/Watkins. Spatial Stochastic Processes: A Festschrift in Honor of Ted Harris on his 70th Birthday
20 Haagerup/Hoffmann-Jørgensen/Nielsen. Probability in Banach Spaces 6: Proceedings of the Sixth International Conference, Sandbjerg, Denmark 1986
21 Eberlein/Kuelbs/Marcus. Probability in Banach Spaces 7: Proceedings of the Seventh International Congress
22 Pinsky. Diffusion Processes and Related Problems in Analysis, Vol. I: Diffusions in Analysis and Geometry
23 Hahn/Mason/Weiner. Sums, Trimmed Sums and Extremes
24 Çinlar/Chung. Seminar on Stochastic Processes, 1990
25 Cambanis/Samorodnitsky/Taqqu. Stable Processes and Related Topics
26 Cruzeiro/Zambrini. Stochastic Analysis and Applications: Proceedings of the 1989 Lisbon Conference
27 Pinsky/Wihstutz. Diffusion Processes and Related Problems in Analysis, Vol. II: Stochastic Flows
28 Durrett/Kesten. Random Walks, Brownian Motion and Interacting Particle Systems

29 ÇINLAR/CHUNG/SHARPE. Seminar on Stochastic Processes, 1991
30 DUDLEY/HAHN/KUELBS. Probability in Banach Spaces 8, 1992
31 KÖREZLIOĞLU/ÜSTÜNEL. Stochastic Analysis and Related Topics, 1992
32 NUALART/SOLE. Barcelona Seminar on Stochastic Analysis: St. Feliu de Guixols, 1991
33 ÇINLAR/CHUNG/SHARPE. Seminar on Stochastic Processes, 1992
34 FRIEDLIN. The Dynkin Festschrift: Markov Processes and their Applications, 1994
35 HOFFMAN-JØRGENSEN/KUELBS/MARCUS. Probability in Banach Spaces 9, 1993
36 BOLTHAUSEN/DOZZI/RUSSO. Seminar on Stochastic Analysis, Random Fields and Applications, 1993
37 BANDT/GRAF/ZÄHLE. Fractal Geometry and Stochastics, 1995
38 KÖREZLIOĞLU/ØKSENDAL/ÜSTÜNEL. Stochastic Analysis and Related Topics V: The Silivri Workshop, 1994
39 ADLER/MÜLLER/ROZOVSKII. Stochastic Modelling in Physical Oceanography
40 CHAUVIN/COHEN/ROUAULT. Trees: Workshop in Versailles, June 14–16, 1995
41 BOVIER/PICCO. Mathematical Aspects of Spin Glasses and Neural Networks.
42 DECREUSEFOND/GJERDE/ØKSENDAL/ÜSTÜNEL. Stochastic Analysis and Related Topics VI: The Geilo Workshop, 1996
43 EBERLEIN/HAHN/TALAGARAND. High Dimensional Probability
44 BRAMSON/DURRETT. Perplexing Problems in Probability: Festschrift in Honor of Harry Kesten
45 DALANG/DOZZI/RUSSO. Seminar on Stochastic Analysis, Random Fields and Applications, 1996
46 BANDT/GRAF/ZÄHLE. Fractal Geometry and Stochastics II
47 GINÉ/MASON/WELLNER. High Dimensional Probability II
48 DECREUSEFOND/ØKSENDAL/ÜSTÜNEL. Stochastic Analysis and Related Topics VII: The Silivri Workshop
49 IMKELLER/VON STORCH. Stochastic Climate Models
50 CRUZEIRO/ZAMBRINI. Stochastic Analysis and Mathematical Physics